SpringerBriefs in Systems Biology

More information about this series at http://www.springer.com/series/10426

Kareem A. Mosa • Ahmed Ismail
Mohamed Helmy

Plant Stress Tolerance

An Integrated Omics Approach

 Springer

Kareem A. Mosa
Department of Applied Biology,
  College of Sciences
University of Sharjah
Sharjah, UAE

Department of Biotechnology, Faculty
  of Agriculture
Al-Azhar University
Cairo, Egypt

Mohamed Helmy
The Donnelly Centre for Cellular
  and Bimolecular Research
University of Toronto
Toronto, ON, Canada

Ahmed Ismail
Department of Biotechnology, Faculty
  of Agriculture
Al-Azhar University
Cairo, Egypt

ISSN 2193-4746   ISSN 2193-4754   (electronic)
SpringerBriefs in Systems Biology
ISBN 978-3-319-59377-7   ISBN 978-3-319-59379-1   (eBook)
DOI 10.1007/978-3-319-59379-1

Library of Congress Control Number: 2017943691

Printed on acid-free paper

This Springer imprint is published by Springer Nature
The registered company is Springer International Publishing AG
The registered company address is: Gewerbestrasse 11, 6330 Cham, Switzerland

Preface

The initial idea for writing this book was formulated during a discussion between us (the book authors) supported by the positive feedback of Dr. Helmy who had already published a book in the "SpringerBriefs in Systems Biology" series. During this discussion, Dr. Helmy shared his experience and introduced the concept of the "SpringerBriefs" series. We then outlined the book proposal and submitted it to the Springer editor Noreen Henson who encouraged and invited us to start working on writing this book for publication in the "SpringerBriefs in Systems Biology" series.

According to the UN DESA report, "World Population Prospects: The 2015 Revision," the world population is expected to reach 8.5 billion and 9.7 billion by the years 2030 and 2050 respectively (http://www.un.org/en/development/desa/news/population/2015-report.html). Therefore, the production of agricultural products may need to be doubled by 2050 in order to meet the required demand for food as a result of the increased population. One of the major focuses in agricultural production research is to understand how plants tolerate unfavorable biotic and abiotic stresses which, therefore, increases the crop yield. In the last two decades, "omics" technologies (such as genomics, transcriptomics, proteomics, and metabolomics) represented a landmark in the development of biological sciences, including plant sciences with novel applications in investigating and improving stress tolerance in plants. This book is specifically tailored to meet the needs of a broader audience, particularly to include postgraduate students and junior researchers in areas such as plant biotechnology, plant omics, system biology, environmental stresses, and bioinformatics. The book is aimed to provide a simple and brief overview of cutting-edge research on "omics" applications in the field of plant sciences, with special focus on different approaches towards plant stress tolerance (including both biotic and abiotic stresses).

The book is divided into four chapters: Chap. 1 introduces the different types of plant stresses (biotic and abiotic) and how plants respond and deal with these stresses. Furthermore, Chap. 2 introduces the omics technologies used to study plant stresses and the bioinformatics platforms associated with these technologies. Chapter 3 provides a general overview of the omics technologies employed to understand biotic stresses with special focus on plant parasitic nematodes as a case study. Lastly, Chap. 4 discusses different omics technologies utilized with functional genomics approaches to study abiotic stress tolerance mechanisms in plants. In conclusion, this book provides an overview of up to date information for graduate students, junior academic scientists, and researchers on utilizing recent advances in omics technologies in the area of plant stresses.

Sharjah, UAE   Kareem A. Mosa
Cairo, Egypt   Ahmed Ismail
Toronto, ON, Canada   Mohamed Helmy

Acknowledgments

We must emphatically express our deepest gratitude to the Department of Biotechnology, Al-Azhar University. Special thanks to the godfather of the department, the man who has most influenced and inspired our undergraduate education at Al-Azhar University, Prof. Fawzy El-feky. We are very grateful to Noreen Henson and Patrick Carr from Springer for their great efforts in pushing this project forward and in facilitating the publication of this book. We would also like to express our great thanks to Dr. Suhail Asrar for language editing this whole book. KAM would like to thank the University of Sharjah, UAE, for administrative support. We also thank our families for their patience and support during the preparation of the book.

Kareem A. Mosa, Ph.D., Sharjah, UAE

Ahmed Ismail, Ph.D., Cairo, Egypt

Mohamed Helmy, Ph.D., Toronto, ON, Canada

April 8, 2017

Contents

Chapter 1
Introduction to Plant Stresses

Abstract Plant stress is a state where the plant is growing in non-ideal growth conditions that increase the demands made upon it. The effects of stress can lead to deficiencies in growth, crop yields, permanent damage or death if the stress exceeds the plant tolerance limits. Plant stress factors are mainly categorized into two main groups; abiotic factors and biotic factors. The abiotic factors include the different environmental factors that affect plant growth (such as light, water, and temperature), while the biotic factors are the other organisms that share the environment and interact with the plants (such as pathogens and pests). Response to stress usually involves complex molecular mechanisms, including changes in gene expression and regulatory networks. In this chapter, we will provide a general overview of the different types of plant stresses, their effects and how plants respond these different types of stress.

Keywords Plant stress • Biotic stress • Abiotic stress • Stress factors • Stress responses • Stress effects • Omics

1.1 Introduction

Plants are in the bottom of the food chain with their unique abilities to produce raw food material. They are the main source of food for the majority of the world population as well as the source of food for animals used for meat or milk production. Further, plants represent a renewable resource for raw materials used in industry. Due to their major importance, early communities situated around water resources that helped them grow their crops. Historical evidence has also shown that nations with abundant food production had faster rates of progress and development (Shao et al. 2009). Climate and environmental factors were major determinants of the types of crops to be planted as well as the crop yields. Consequently, they determine the geographical distribution of plants (Bressan et al. 2009). Therefore, most crops are now grown in suboptimal environments that maximize the production and minimize the cost of adaptation (Atkinson and Urwin 2012).

© The Author(s) 2017
K.A. Mosa et al., *Plant Stress Tolerance*, SpringerBriefs in Systems Biology,
DOI 10.1007/978-3-319-59379-1_1

Since environmental factors play a crucial role in the growth, development and productivity of crop yields in plants, unfavorable changes in the environmental factors can result in deficiencies in plant growth, decreases in crop yields, permanent damage, and even death (Duque et al. 2013). Unlike animals, plants cannot respond to the changes in their environment by moving to a more suitable environment, a feature that helps in surviving seasonal weather and resource changes for instance. Therefore, plants need to tolerate the unfavorable environmental conditions in order to minimize their impact (Boyer 1982). A combination of processes and factors determines the ability of plants to tolerate the unfavorable conditions, including a wide spectrum of molecular processes and regulatory mechanisms (Duque et al. 2013).

Another factor that affects plant growth, reproduction, distribution and crop yields is the other living organisms that share the same environment and interact with the plant. Microorganisms, insects, weeds and other wild or cultivated plants can cause damages to the plant in several ways, resulting in losing part or all the crop yields or reducing its quality (Fig. 1.1) (Peterson and Higley 2000). Thus, understanding how plants tolerate unfavorable environmental or biological conditions is a major focus in agricultural research due to its economical impact. In general, plant exposure to unfavorable conditions, either environmental or biological, is known as "Plant Stress".

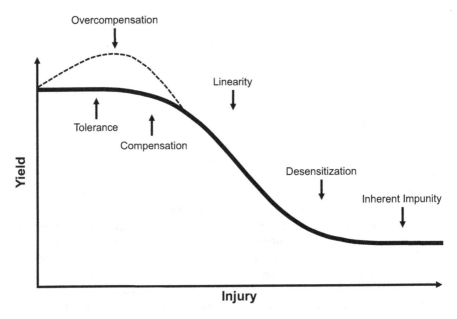

Fig. 1.1 The damage curve that describes the theoretical relationship between injury/stress and yield. The curve contains six named segments that indicate unique responses to injuries/stresses. (1) Tolerance: no loss of yields, (2) Overcompensation: increased yields, (3) Compensation: decreased yields, (4) Linearity: decreased yields with higher rate, (5) Desensitization: decreased yields with lower rate, and (6) Inherent Impunity: yields with injury are less than yields with no injury at a constant slope, adapted from Peterson and Higley (2000)

1.2 Definitions of Plant Stress

The stress term in biology was initially derived from the physics and mechanics term "Stress", which was used to describe the amount of force in a given unit area (Wardlaw 1972). The similarity came from the fact that if we apply certain stress on an elastic material the effect will be revisable, while applying stress on a plastic material will cause irreversible changes that can lead to the breaking of the material. In biology, the exposure to unsuitable environmental conditions, such as high temperature, can be within certain limits that the plant response can reverse or minimize its effect and, therefore, restore its normal growth or production rates. This process is very similar to applying stress or pressure on an elastic material. However, the exposure to unfavorable conditions with high dosage and/or for long duration can exceed the limits of the plant responses and cause permanent damage or death. This is very similar to applying stress or pressure on a plastic or hard material that results in bending or breaking (Kranner et al. 2010).

In 1987, Walter Larcher defined plant stress as "a state in which increasing demands made upon a plant lead to an initial destabilization of functions, followed by normalization and improved resistance. If the limits of tolerance are exceeded and the adaptive capacity is overtaxed, permanent damage or even death may result" (Larcher 1987). According to this definition, plant stress is a state where the plant experiences a change that demands response. If the demand is within the plant tolerance limits, the plant response can restore the normal state. Therefore, the stress can be considered as a temporary state. However, if the stress exceeds the tolerance capacity of the plant, a permanent damage will occur. The degree of the damage is relative to the stress strength and duration. It can even lead to death if the plant responses fail to deal properly with the stress (Pahlich 1993).

There are several other definitions of stress that were formulated later by plant scientists. For instance, stress was defined as "any unfavorable condition or substance that affects or blocks a plant's metabolism, growth or development" (Lichtenthaler 1996). Alternatively, it was also described as "a condition caused by factors that tend to alter an equilibrium" (Gaspar et al. 2002; Kranner et al. 2010) and as "changes in physiology that occur when species are exposed to extraordinarily unfavorable conditions that need not represent a threat to life but will induce an alarm response" (Gaspar et al. 2002). Further, it can be defined as "an external factor that exerts a disadvantageous influence on the plant" (Taiz and Zeiger 1991) and more generally as "a significant deviation of the optimal condition of life" (Lichtenthaler 1998). Despite the differences between the several definitions of plant stress, they are all centered on describing a change in the conditions that affect the plant, the plant response to this change and the level of damage (if any) that can be brought about by the change. One reason why various scientists define stress using different definitions is that each of them was studying different type(s) of plant stresses.

1.3 Types of Plant Stresses

Plant stress can be classified into several types in accordance with numerous factors as follows:

1. According to the type of factors that cause the stress: Plant stresses can be classified into "Abiotic Stresses" that are caused by non-living factors (such as drought, changes in temperatures and salinity), and "Biotic Stresses" that are caused by living organisms (such as microorganisms, insects and other plants) (Kranner et al. 2010).
2. According to the effect of the stress: Plant stress can be classified into stresses with positive effects, or "Eustress", and stresses with negative effects, or "Distress" (Kranner et al. 2010). The balance between tolerance and sensitivity determines the effect of the stress. Water deficiency can cause hardening as long as it is below the permanent wilting point "positive", and can be lethal if the deficiency reaches the permanent wilting point "negative" (Kranner et al. 2010).
3. According to the persistence of the stress: Plant stress can be classified into "Short-term Stresses", where the plant can overcome the stress by means of adaptation, acclimation and repair mechanisms, and persistent or "Long-term Stresses" that result in substantial and irreversible damages (Lichtenthaler 1996; Kranner et al. 2010).
4. Plant Stresses can also be classified into "Internal Stresses" that come from within the plant and "External Stresses" that exist outside the plant. External and internal stresses are usually referred to as stress factors and stresses, respectively (Kranner et al. 2010).

Since the focus of this book is biotic and abiotic stresses, we will describe them in more details.

1.3.1 Biotic Stress

Biotic stress is a result of interactions between the plant and another living organism(s) that results in either partial damage that the plant can overcome or significant damages that the plant cannot survive. Almost all types of living organisms can cause biotic stresses including pathogenic bacteria, fungi or viruses as well as nematodes and insects (Kranner et al. 2010). Animals and plants (wild or cultivated) are also causes of biotic stresses. However, biotic stresses caused by microorganisms are mostly in the form of diseases (bacterial, fungal and viral plant diseases) or parasitism. Insects and animals cause physical damages mostly through perdition, while plants cause stresses through competition and phyto-parasitism.

According to the above classifications, biotic stress is an external biological stress that affects the plant. Its impact can affect all the organization levels of the plant, including molecules, organelles, cells, tissues, organs, whole plants or even whole plant populations. The impact can influence some or all these levels and responses can be noticeably different from one level to another. Therefore, understanding the effects and the stress response mechanisms requires physiological, cellular and molecular measures at all these levels (Peterson et al. 2001). In Chap. 3, biotic stresses and plant responses to forms of biotic stresses will be discussed in more detail.

1.3.2   Abiotic Stress

Each plant has optimal growth conditions that help it to reach the production stage as well as completing its life cycle through reproduction. Abiotic stresses are the result of changes in non-biological factors, mostly environmental or nutritional, that affects the plant's growth, reproduction or life. Therefore, they demand responses from the plant that help to restore normal conditions or allow the plant to minimize the detrimental effects of these changes (Shao et al. 2009). Abnormalities in water status, nutrient levels, wind conditions, temperature, and salinity are some of the sources of abiotic stresses (Kranner et al. 2010). Thus, abiotic stresses are physical stresses that can be easily measured through the utilization of physical terms (Peterson et al. 2001).

According to the above classifications, abiotic stresses are external environmental and nutritional stresses and can affect the plant for short or long durations. Therefore, they can have either positive effects (termed as "Eustress") or negative effects (termed as "Distress") (Kranner et al. 2010). We mentioned above the dual effect of water deficiency that changes in accordance with the deficit amount, by either reaching the permanent wilting point or not. Another important example is the temperature. While extreme temperature can be lethal (Harding et al. 1990), extreme heat and cold can induce hardening (Somersalo and Krause 1989) and can be used alternatively to induce seed dormancy (Finch-Savage and Leubner-Metzger 2006). In Chap. 4, abiotic stresses and plant responses to forms of abiotic stresses will be discussed in more detail.

1.4   Plant Stress Factors

As mentioned above, the major determinant of plant growth, production and reproduction is growth in optimal conditions. However, plants are exposed to a wide spectrum of biotic and abiotic stress factors, also known as stressors, which deviate the growth conditions from the optimal point. Plant exposure to such a stressor(s) triggers certain response mechanisms for adaptation and defense based on highly developed metabolic, signaling and regulatory networks as described in the following sections (Lichtenthaler 1998; Sazzad 2007). In this section, we will briefly overview some of the major abiotic and biotic stressors and their effects on plants.

1.4.1   Drought

Drought or water deficit is one of the major abiotic stressors that directly affects all aspects of plant growth and development causing dehydration and resulting in reduced crop yields. The main result of drought shock is a metabolic and osmotic imbalance leading to turgor loss and stomata closure (Zhu 2001; Sazzad 2007). This limits the carbon dioxide uptake and, therefore, causes cell growth repression and photosynthesis reduction (Shinozaki and Yamaguchi-Shinozaki 2000). If the water deficit stress is prolonged enough to reach the permanent wilting point, it will lead to plant death.

1.4.2   Temperature

Temperature is another major stressor that has effects on the plant starting from seed germination to reproduction (Kranner et al. 2010). Changes in the surrounding temperature can result in unmanageable stresses that lead to permanent damage or death. Furthermore, since each species has an optimal growth temperature, zone temperature is a major factor in the determination of plant species distribution on the planet. The range of daytime temperatures that plants can experience is between −70 and 60 °C. Local topology as well as altitude significantly affects the area temperature (Sazzad 2007).

Temperature can cause stress to plants via two different means; cold and heat. Extreme drops in temperature below optimal growth degrees causes severe mechanical and physical damages to the plants. When the temperature falls below zero, ice formation starts in the intercellular spaces before the intracellular fluids due to the lower freezing point of the latter. Therefore, this results in pressure to the cell walls and membranes resulting in severe cell disruption (Olien and Smith 1977; Sazzad 2007). Another consequence of freezing due to low temperatures is freezing injuries. Freezing injuries result from the cell desiccation that takes place when the unfrozen fluid or water in cells moves to the intercellular spaces. This happens due to the increased water potential outside the cell caused by ice formation (Thomashow 1999; Sazzad 2007). At the molecular level, cold causes membrane damage, cell lysis, molecular precipitation and ROS production (Sazzad 2007; Thomashow 1998, 1999; Pearce 1999).

Plant exposure to high temperatures slows down growth and reduces the nutritional uptake rate, while exposure to extremely high temperatures can result in heat shock. Extremely high temperatures increase the rate of evaporation from the stomata causing wilting, with further prolongation resulting in permanent damage to the plant or even death. Further, extremely high temperatures affect the vitality of different organs, especially those involved in reproduction and pollination. It was shown that extreme temperature has a significant negative impact on pollination as it reduces the viability of both pollen and silk, which limits complete fertilization. At the molecular level, heat shock causes the expression of otherwise unexpressed proteins, modifying the DNA transposition frequency and causing protein denaturation (Peterson and Higley 2000).

1.4.3   Salinity

Despite the great importance of the availability of minerals and nutrients in soil, the unnecessary presence of soluble salts causes severe osmotic and ionic stress in plants (Sazzad 2007). Salt stress or salinity is a result of the presence of excessive amounts of water-soluble salts such as sodium sulphate (Na_2SO_4), sodium nitrate ($NaNO_3$), sodium chloride ($NaCl$), sodium carbonates ($NaHCO_3$ and Na_2CO_3), potassium sulphate (K_2SO_4), calcium sulfate ($CaSO_4$), magnesium sulphate ($MgSO_4$) and magnesium chloride ($MgCl_2$) (Flowers et al. 1977; Sazzad 2007). Most of these salts are important to the plant and help in improving its growth and metabolism. However, they may turn toxic if present in excessive concentrations (Sazzad 2007). For instance, the optimal concentration of sodium chloride enhances plant growth, but higher concentrations have very harmful inhibitory effects on seed germination and seedlings in salt-prone soils as well as causes severe damage to plant growth and development (Tester and Davenport 2003; Shabala et al. 2015).

In general, plants can be classified into two main groups according to their ability to grow under high salinity conditions as either halophytes or glycophytes. Halophytes are plants that can tolerate high salinity levels similar to seawater or even higher levels in some cases (Zhu 2001; Sazzad 2007). Halophytes use two main approaches to avoid the harmful effects of high salinity, i.e. limiting the salt uptake and reducing the salt concentration in the cytoplasm and cell wall (Munns et al. 2006). Glycophytes or salt-sensitive plants cannot tolerate high salinity levels and are unable to utilize the two approaches of minimizing high salinity effects that are employed by the halophytes. Therefore, they accumulate toxic levels of salt in the cytosol (Munns et al. 2006; Sazzad 2007).

Salinity mainly affects the underground organs of the plant, such as roots and tuber, as well as affecting seed germination and seedling viability (Zhu 2001; Kranner et al. 2010). The extreme osmotic and ionic stress around the roots results in a reduction of the plant's water uptake capacity, which reduces growth and metabolism rates (Flowers et al. 1977; Sazzad 2007). The prolonged salt stress causes stomata closure reducing the carbon dioxide uptake capacity. This salt toxicity results in reduced photosynthetic capacity, leading to senescence and, consequently, death as the plant fails to sustain a proper growth rate (Flowers et al. 1977; Zhu 2001; Sazzad 2007).

1.4.4 Hypoxia and Anoxia

The availability of free water to plants is indispensable. However, excessive amounts of water in the area surrounding the roots can be harmful, suffocating or even lethal (Sazzad 2007). Water stress suffocates roots by blocking the free oxygen transfer between the soil and atmosphere (Drew 1997). Waterlogged soils are always a result of flooding, heavy rains or ice and snow melting during and after winter. Such soils are usually known to lack or possess limited free oxygen due to the reduction of gas exchange (Jackson and Colmer 2005). Under hypoxic conditions, plants use alternative carbohydrate consumption and anaerobic metabolism as well as alternative organs for oxygen and gas exchange, e.g. leaves or aerial roots (Fukao and Bailey-Serres 2004).

Hypoxia and anoxia are the partial and complete absence of free oxygen from the soil, respectively. Both hypoxia and anoxia largely affect growth, development and crop yields (Vartapetian et al. 2003). Similar to drought stress, hypoxia causes stomata closure and wilting as a defense mechanism utilized by the plant to reduce the damaging effects of oxygen insufficiency. It also reduces photosynthetic, metabolic and transpiration rates (Crawford and Braendle 1996). One of the major consequences of hypoxia is the generation of reactive oxygen species (ROS) in plant tissues due to initial oxygen insufficiency followed by oxygen availability (Blokhina et al. 2003; Farnese et al. 2016).

1.4.5 Reactive Oxygen Species (ROS)

Reactive oxygen species (ROS) are molecules that are formed as a result of different types of stresses. ROS can be formed under salinity, drought, heat shock, hypoxia and oxidative stress conditions. Hydrogen peroxide (H_2O_2), superoxide anion (O_2^-) and hydroxyl ion (OH^-) are some famous ROS molecules that cause membrane and macromolecular damage (Blokhina et al. 2003; Sazzad 2007; Farnese et al. 2016). To reduce the damaging effects of ROS toxic components, plants utilize several antioxidant defense approaches in order to scavenge them and increase plant tolerance to different stress factors. Primarily, plant cells use several antioxidants to induce enzymes such as CAT, superoxide dismutase (SOD), APX, glutathione reductase and non-enzyme molecules, such as ascorbate, glutathione, carotenoids and anthocyanins (Mittler 2002; Gould et al. 2002; Blokhina et al. 2003; Sazzad 2007). In addition, several other molecules demonstrate antioxidant effects and function as ROS scavengers such as proteins and amphiphilic molecules (Noctor and Foyer 1998; Gould et al. 2002; Blokhina et al. 2003; Sazzad 2007; Farnese et al. 2016).

1.4.6 Light

Sunlight is an essential factor for plant energy and metabolite production via photosynthesis. Plants evolve in a way that allows them to cope with fluctuations in light exposure. Light exposure periods fluctuate dynamically with different intervals. Therefore, plants develop mechanisms to maximize the utility of the available light under low irradiance conditions and other mechanisms that allow them to avoid the damaging effects of prolonged light exposure (Adamiec et al. 2008). Obviously, the limited availability of light or solar energy reduces plant energy and metabolite production through slowing down photosynthesis processes and leading to slower growth rates and decrease in crop yields. Prolonged exposure to light or solar energy causes a potential risk of photo-damage and increased ROS production (Barta et al. 2004; Adamiec et al. 2008).

1.4.7 Wounds

Insect herbivore infestation, as well as birds and animals, cause wounds or tissue damage to plants. Besides their damaging effects, small wounds are also considered hindrances to the solid defense system of plants as they allow secondary infections by pathogens which are not equipped with mechanical penetration organs, such as bacteria and fungi (Sazzad 2007). Therefore, plants in many cases respond to wounds in a similar manner to pathogen attacks in order to reduce the chances of secondary infections, e.g. the plant response to small foliage wounds caused by phloem-feeding whiteflies and aphids (Walling 2000). Bigger wounds and extensive tissue damage caused by chewing insects (such as beetles and caterpillars) or cell content feeders (such as mites and thrips) are treated differently by the plant. The insect feeding itself initiates wound-induced responses, while the feeding damage can induce extensive responses due to the high content of stimulating oral secretions (Walling 2000; Denekamp and Smeekens 2003; Sazzad 2007; Hettenhausen et al. 2015).

1.4.8 Pathogens

Pathogens are parasitic organisms that cause diseases to plants; this includes a wide spectrum of microorganisms (e.g. bacteria, fungi, and viruses), protozoa and nematodes. Pathogenic diseases all over the world cause significant crop losses every year (Baker et al. 1997; Agrios 2005; Sazzad 2007). Parasitic organisms usually utilize the host plant for feeding, growing, multiplying and sheltering, causing significant damages to the host that can lead to death. The damage caused by the pathogens is mainly through the acquisition of nutrients from the plant host, feeding on the host's different organs and causing physical damage such as small and large wounds (Agrios 2005). Furthermore, pathogens cause significant harmful effects through the secretion of toxic substances that lead to direct damage of the host plant tissues or an induced stress response by the host plant. This pathogen-host interaction is a battle at the biochemical and molecular levels that leads to several functional metabolic and physiological changes resulting in growth and shape disorders, impaired production or even the death of the host (Baker et al. 1997; Sazzad 2007).

1.4.9 Competition

Plants compete for resources with neighboring plants, which creates a special type of stress called "Competition stress". Competition stress can be within the same species, other cultivated species or wild plants. For instance, a high seeding rate can result in a high population density of the same species that is unsuitable based on the available resources or soil type. Therefore, a competition stress occurs for moisture, space, light, and soil nutrients (Peterson and Higley 2000). The competition stress intensity is relative to the population density and resource availability. Indeed, this type of stress results in reduced crop yields and/or fruit quality and can lead to impaired growth or even be lethal if the population density is very high.

Similar to competing with the same species, plants can also come under competition stress with other species especially weeds, which is referred to as "weed competition". Estimations show that weed competition is responsible for 10–15% of crop yields losses worldwide (Froud-Williams 2003). Unlike competition with the same species, weed competition is unrelated to the seeding rate but is highly related to pre-seasonal soil treatments and field preparations as well as weed-control programs and other approaches applied during the season. Inappropriate preparation or using less-than-effective herbicide programs places the plants under weed competition and risks the crop yields (Peterson and Higley 2000; Lehoczky and Reisinger 2003).

1.5   Plant Stress Response Phases

When one or more stresses changes the plant optimal conditions, the plant uses a special mechanism to detect this change called "Stress sensing". There are several stress sensing mechanisms that are used by plants depending on the species, organ and stress type (Kranner et al. 2010). For instance, wind and light stresses affect the aboveground parts of the plant, while salinity and drought affect the underground parts of plant, triggering distinct stress sensing mechanisms in each case. The most common stress sending models for chemical and radiation stress are the receptor-substrate and receptor-photon binding models (Verslues and Zhu 2005). The water concentration in the soil can be sensed using osmosensors in the root cells, while the sugar generation process can be used in sensing stresses that affect signaling and development (Urao 1999; Rolland et al. 2006).

Stress sensing is the very first event that appears after exposure to stressors and triggers the plant stress response(s). There are four main phases of plant stress and response to stress events. The phases appear to be based on the duration and intensity of the stressor (Lichtenthaler 1998; Kranner et al. 2010). Here, we will briefly introduce each of the four phases according to one of the most acceptable plant-response models (Fig. 1.2) (Lichtenthaler 1998).

1.5.1   Alarm Phase

The alarm phase is the stress reaction that takes place based on the recognized change in the optimal growth conditions. It can be distinguished with some main characteristics such as the functional deviation in comparison with normal conditions, declining vitality and a higher rate of catabolism (Lichtenthaler 1998). In most cases, stressors act together so that several related stressors can trigger the alarm phase simultaneously. For instance, in hot summer, plants are concurrently exposed to high temperature, water deficiency and high-light stresses that result in the triggering of stress responses (Kranner et al. 2010). In this phase, plants activate stress coping mechanisms and thus only those with low or no stress tolerance mechanisms can suffer from acute damage (Fig. 1.2) (Lichtenthaler 1998).

1.5.2   Resistance Phase

In the resistance phase, also known as the restitution phase, the plant responds to stressors still affecting its growth and reproduction through adaptation, repair and hardening processes (Lichtenthaler 1998). The stress tolerance mechanisms

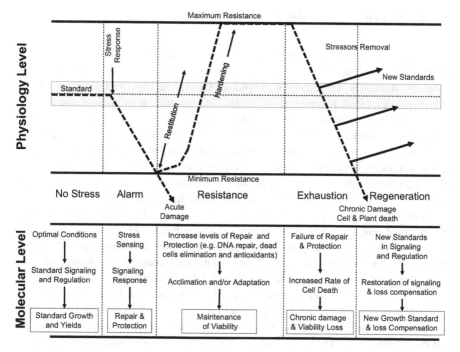

Fig. 1.2 Plant stress response phases. The figure describes the general concepts of the phases and phase sequences induced by the plant in response to stressors at the physiological level (*top panel*) and molecular level (*lower panel*), adapted from Lichtenthaler (1996) and Kranner et al. (2010)

activated in the alarm phase as well as the adaptation, repair and hardening processes results in the establishment of a new physiological standard that is optimal for the stress conditions. However, each plant has a resistance maximum, which cannot be exceeded. If the stressor has a prolonged effect and/or the stress dosage is higher than the plant resistance can handle, the plant enters the next phase (Fig. 1.2) (Lichtenthaler 1998).

1.5.3   Exhaustion Phase

Plants enter the exhaustion phase, also known as the end phase, if they are exposed to a prolonged stressor and/or an overloading stress-dose that the stress coping mechanisms fail to handle (Lichtenthaler 1998). Such prolonged or overloading stress results in progressive loss of vitality. Continuation of the same stress conditions will result in severe damage and, ultimately, death. The damage rate is also dependent on the species, organ, time and dosage (Fig. 1.2) (Lichtenthaler 1998).

1.5.4 Regeneration Phase

If the stressor(s) that affects the plant is removed before the domination of the permanent senescence process, the plant can regenerate and restore itself to physiological standards by entering a phase known as the regeneration phase (Lichtenthaler 1998; Kranner et al. 2010). The restoration to physiological standards will allow the plant to survive. However, any permanent damage caused by prolonged or overloading stress will not be recovered. The new physiological standards will be somewhere between the resistance minimum and resistance maximum depending on the time and stage of exhaustion when the stressor was removed (Fig. 1.2) (Lichtenthaler 1998). Therefore, it is expected the plant can survive but with limited growth (e.g. will be a dwarf) or reproduction abilities.

Despite the response phase, responses to stressors usually involve physiological, biochemical and molecular level actions (Lichtenthaler 1998; Shao et al. 2009; Kranner et al. 2010). Taking responses to drought stress as an example, the plant physiological responses in this case include root signals recognition, osmotic adjustment, leaf water potential, reduced internal CO_2 concentration, photosynthesis decline and a reduced overall growth rate. At the biochemical level, responses can involve decreased photochemical efficiency, stress metabolites accumulation, antioxidative enzymes increase and reduction in reactive oxygen radicals or species (ROS) accumulation. The molecular response to salinity can involve the expression of stress responsive and abscisic acid (ABA) genes, synthesis of specific proteins, drought stress tolerance and differential expression levels of genes related to several metabolic pathways (Reddy et al. 2004; Shao et al. 2007, 2009).

1.6 Plant Stress Measurement

As mentioned in the above sections, plants under stress conditions experience changes in morphological and structural characteristics as well as biochemical alterations. At the molecular level, plants experience changes in gene expression levels, protein expression levels and metabolite levels. These changes are mainly to cope with the stress conditions and to tolerate the absence or excessive levels of nutrients (Peterson and Higley 2000; Kranner ct al. 2010; Duque et al. 2013). The investigation of stress response and tolerance requires taking accurate measurements of the altered characteristics and expression levels to compare between the normal and stress conditions. This step is indispensable to understand stress effects and tolerance mechanisms as well as estimate the upper and lower limits of stress tolerance to certain stressor(s) (Kruse et al. 2011; Minina et al. 2013).

Plant stress measurements are mainly categorized into three main types of measurements. The first type assesses measurements from living plants by primarily estimating the plant viability. The viability of plants under stress is compared with the average viability (e.g. growth, production and reproduction) of plants growing under normal conditions (Fig. 1.3) (Buchner et al. 2013). The second type of measurements focuses on measuring changes in a particular biological process such as photosynthesis or cell

Fig. 1.3 The generalized crop-yield responses to insect injury, adapted from Peterson and Higley (2000)

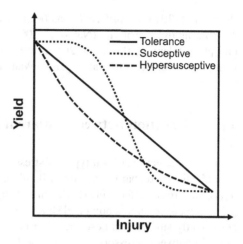

signaling in response to the stressor (Bunce 2009). The last type of measurements involves measuring the plant response to stress at the molecular level by measuring the changes in the expression levels of genes and proteins (Fan et al. 2013; Singh and Jwa 2013; Wang et al. 2013a), the changes in protein phosphorylation (Vialaret et al. 2014) or changes in metabolism (Saruyama et al. 2013).

1.7 Plant Stress and Evolution

Stresses, especially abiotic, are major drivers of evolution in plants. This is mainly due to the extreme selection pressure the stress conditions impose on plants, making them face an outstanding survival challenge (Jackson et al. 2008). Temperature, water and soil composition are the three main determinants of the distribution of plant species worldwide due to their direct effects on plant life and their exceptional stress capacity. Each of them can act on plants through two opposite types of extreme stress, which is drought or flooding for water, extreme heat or cold for temperature and absence or overabundance of nutrients in soil (Amtmann et al. 2005; Jackson et al. 2008). Under such extreme pressure, only plant species with adequate stress tolerance mechanisms can survive. Therefore, plants develop stress tolerance mechanisms that help in surviving extreme conditions throughout the long evolutionary process [this topic was intensively reviewed by Beerling (2007)].

In order to investigate the stress tolerance mechanisms developed by plants during evolution, plant biologists searched for a model organism for extreme condition (or "extremophile") research (Amtmann et al. 2005). *Thellungiella halophila*, a close relative of the universal plant model organism *Arabidopsis thaliana*, was proposed as a plant extremophile model organism since it tolerates extreme cold, drought, and salinity (Bressan et al. 2001; Inan et al. 2004; Taji et al. 2004). *T. halophila* helped in the discovery and better understanding of several mechanisms of stress tolerance and adaptation as well as in investigating the relationship between stress and evolution

(Song et al. 2013). *T. halophila* has been used to study salt tolerance (Wang et al. 2013b), salt oversensitivity (Nah et al. 2009), boron tolerance (Lamdan et al. 2012), phosphate tolerance (Pei et al. 2012), water stress (Arbona et al. 2010) and in the identification of stress tolerance genes (Wang et al. 2010).

1.8   Interaction Between Different Stresses

Plants are subject to several types of stress at the same time, which makes stressor identification and characterization difficult and limits the ability to understand and treat each stressor effectively (Peterson and Higley 2000). However, several research projects focused on this point and showed that the relation between stressors can be consistently studied and described. Furthermore, studies have shown that the relationship between stressors is one of the factors that affects the plant response to stresses (Higley et al. 1993). It is common for several stressors to affect the plant at the same time. For instance, summer seasons are hot and dry with prolonged daytime. In this situation, plants can be under heat stress, drought stress and high-light stress concurrently.

The interaction between stresses means that one of them depends on the other or that the effect of one of them is dependent on the other. For instance, the effect of wound stress caused by an insect depends on the plant age, injured organ, environmental conditions and the existence of another organism that can utilize the wound to cause another type of stress (Lichtenthaler 1998; Peterson and Higley 2000; Atkinson and Urwin 2012). In general, plant stressor interactions can take two main forms. The first instance is when the occurrence of the second stressor changes the plant response to the first stressor. The second is when the occurrence of a particular stressor alters the incidence of another stressor, e.g. when the first stressor creates favorable conditions for the other as in the case of insect wounds that promote the development of bacterial infection (Higley et al. 1993).

A classification of the types of interaction between stressors was proposed by Higley et al. where interactions were classified into three main types and several subtypes (Higley et al. 1993). The following is a brief overview of the classification put forward by Higley et al. and the definition of each interaction type:

1. Independence: plant responses to each stressor is not affected or influenced by the occurrence of the other.
2. Interaction or Dependence: plant responses to each stressor is affected or influenced by the occurrence of the other.

    (a) Stress Response Interactions: plant responses to multiple stresses are not equal to the sum of responses to the individual stresses. This indicates that the physiological and molecular processes affected by the stresses are interrelated with the damage.

(b) Stress Incidence Interactions: plant responses to the second stress changes due to the existence of the first stress.

The occurrence of an initial stress changes the incidence of a subsequent stress. This model is very common in biotic stresses caused by microorganisms after the plant is affected by another primary stress (e.g. wound chewing insects).

3. False Relationships: failure to identify the nature of the relationship between two or more stressors

(a) False Independence: identifying multiple stressors as dependent when they are in fact independent.
(b) False Interaction: identifying multiple stressors as independent when they are in fact dependent.

Recent molecular biology studies confirm that the plant response to an individual stressor is different than the response to multiple stressors (Atkinson and Urwin 2012). This proves that the effects are non-additive. Instead, the existence of one stressor can have an enhancing or reducing effect on the susceptibility to another stressor. Biotic and abiotic stresses can interact through complex hormonal signal-

Fig. 1.4 The flow of events that take place in response to multiple stressors, e.g. combined biotic and abiotic stresses, adapted from Atkinson and Urwin (2012)

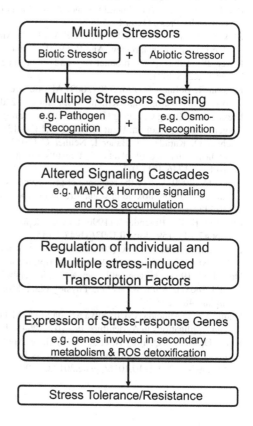

ing pathways that result in synergetic or antagonistic effects against each other (Rizhsky et al. 2004). Figure 1.4 shows the key events involved in the response to multiple stressors at the molecular level.

References

Adamiec M, Drath M, Jackowski G (2008) Redox state of plastoquinone pool regulates expression of Arabidopsis thaliana genes in response to elevated irradiance. Acta Biochim Pol 55:161–173.

Agrios G (2005) Plant Pathology, 5th edn. Elsevier, London

Amtmann A, Bohnert HJ, Bressan RA (2005) Abiotic stress and plant genome evolution. Search for new models. Plant Physiol 138:127–130. doi: 10.1104/pp.105.059972

Arbona V, Argamasilla R, Gómez-Cadenas A (2010) Common and divergent physiological, hormonal and metabolic responses of Arabidopsis thaliana and Thellungiella halophila to water and salt stress. J Plant Physiol 167:1342–1350. doi: 10.1016/j.jplph.2010.05.012

Atkinson NJ, Urwin PE (2012) The interaction of plant biotic and abiotic stresses: from genes to the field. J Exp Bot 63:3523–3543. doi: 10.1093/jxb/ers100

Baker B, Zambryski P, Staskawicz B, Dinesh-Kumar SP (1997) Signaling in plant-microbe interactions. Science 276:726–733.

Barta C, Kálai T, Hideg K, et al (2004) Differences in the ROS-generating efficacy of various ultraviolet wavelengths in detached spinach leaves. Funct Plant Biol 31:23. doi: 10.1071/FP03170

Beerling D (2007) The Emerald Planet: How Plants Changed Earth's History, 1st edn. Oxford University Press, Oxford

Blokhina O, Virolainen E, Fagerstedt K V (2003) Antioxidants, oxidative damage and oxygen deprivation stress: a review. Ann Bot 91:179–194. doi: 10.1093/aob/mcf118

Boyer JS (1982) Plant productivity and environment. Science 218:443–448. doi: 10.1126/science.218.4571.443

Bressan R, Bohnert H, Zhu J-K (2009) Abiotic stress tolerance: from gene discovery in model organisms to crop improvement. Mol Plant 2:1–2. doi: 10.1093/mp/ssn097

Bressan RA, Zhang C, Zhang H, et al (2001) Learning from the Arabidopsis experience. The next gene search paradigm. Plant Physiol 127:1354–1360.

Buchner O, Karadar M, Bauer I, Neuner G (2013) A novel system for in situ determination of heat tolerance of plants: first results on alpine dwarf shrubs. Plant Methods 9:7. doi: 10.1186/1746-4811-9-7

Bunce JA (2009) Use of the response of photosynthesis to oxygen to estimate mesophyll conductance to carbon dioxide in water-stressed soybean leaves. Plant Cell Environ 32:875–81. doi: 10.1111/j.1365-3040.2009.01966.x

Crawford RMM, Braendle R (1996) Oxygen deprivation stress in a changing environment. J Exp Bot 47:145–159. doi: 10.1093/jxb/47.2.145

Denekamp M, Smeekens SC (2003) Integration of wounding and osmotic stress signals determines the expression of the AtMYB102 transcription factor gene. Plant Physiol 132:1415–1423.

Drew MC (1997) Oxygen deficiency and root metabolism: injury and acclimation under hypoxia and anoxia. Annu Rev Plant Physiol Plant Mol Biol 48:223–250. doi: 10.1146/annurev.arplant.48.1.223

Duque AS, de Almeida AM, da Silva AB, da Silva JM, et al (2013) Abiotic Stress—Plant Responses and Applications in Agriculture. doi:10.5772/45842

Fan X-D, Wang J-Q, Yang N, et al (2013) Gene expression profiling of soybean leaves and roots under salt, saline-alkali and drought stress by high-throughput Illumina sequencing. Gene 512:392–402. doi: 10.1016/j.gene.2012.09.100

Farnese FS, Menezes-Silva PE, Gusman GS, Oliveira JA (2016) When bad guys become good ones: the key role of reactive oxygen species and nitric oxide in the plant responses to abiotic stress. Front Plant Sci 7:471. doi: 10.3389/fpls.2016.00471

Finch-Savage WE, Leubner-Metzger G (2006) Seed dormancy and the control of germination. New Phytol 171:501–523. doi: 10.1111/j.1469-8137.2006.01787.x

Flowers TJ, Troke PF, Yeo AR (1977) The mechanism of salt tolerance in halophytes. Annu Rev Plant Physiol 28:89–121. doi: 10.1146/annurev.pp.28.060177.000513

Froud-Williams RJ (2003) Weed Competition. In: Naylor REL (ed) Weed Manag. Handb., 9th edn. Blackwell Science Ltd, Oxford, UK, pp 16–38

Fukao T, Bailey-Serres J (2004) Plant responses to hypoxia—is survival a balancing act? Trends Plant Sci 9:449–456. doi: 10.1016/j.tplants.2004.07.005

Gaspar T, Franck T, Bisbis B, et al (2002) Concepts in plant stress physiology. Application to plant tissue cultures. Plant Growth Regul 37:263–285. doi: 10.1023/A:1020835304842

Gould KS, McKelvie J, Markham KR (2002) Do anthocyanins function as antioxidants in leaves? Imaging of H_2O_2 in red and green leaves after mechanical injury. Plant Cell Environ 25:1261–1269. doi: 10.1046/j.1365-3040.2002.00905.x

Harding SA, Guikema JA, Paulsen GM (1990) Photosynthetic decline from high temperature stress during maturation of wheat: I. Interaction with senescence processes. Plant Physiol 92:648–653. doi: 10.1104/pp.92.3.648

Hettenhausen C, Schuman MC, Wu J (2015) MAPK signaling: a key element in plant defense response to insects. Insect Sci 22:157–164. doi: 10.1111/1744-7917.12128

Higley LG, Browde JA, Higley PM (1993) International Crop Science I. Int Crop Sci I. doi:10.2135/1993.internationalcropscience.c120

Inan G, Zhang Q, Li P, et al (2004) Salt cress. A halophyte and cryophyte Arabidopsis relative model system and its applicability to molecular genetic analyses of growth and development of extremophiles. Plant Physiol 135:1718–1737. doi: 10.1104/pp.104.041723

Jackson MB, Colmer TD (2005) Response and adaptation by plants to flooding stress. Ann Bot 96:501–505. doi: 10.1093/aob/mci205

Jackson MB, Ishizawa K, Ito O (2008) Evolution and mechanisms of plant tolerance to flooding stress. Ann Bot 103:137–142. doi: 10.1093/aob/mcn242

Kranner I, Minibayeva F V, Beckett RP, Seal CE (2010) What is stress? Concepts, definitions and applications in seed science. New Phytol 188:655–673. doi: 10.1111/j.1469-8137.2010.03461.x

Kruse J, Rennenberg H, Adams MA (2011) Steps towards a mechanistic understanding of respiratory temperature responses. New Phytol 189:659–677. doi: 10.1111/j.1469-8137.2010.03576.x

Lamdan NL, Attia Z, Moran N, Moshelion M (2012) The Arabidopsis-related halophyte Thellungiella halophila: boron tolerance via boron complexation with metabolites? Plant Cell Environ 35:735–746. doi: 10.1111/j.1365-3040.2011.02447.x

Larcher W (1987) Stress bei Pflanzen. Naturwissenschaften 74:158–167. doi: 10.1007/BF00372919

Lehoczky E, Reisinger P (2003) Study on the weed-crop competition for nutrients in maize. Commun Agric Appl Biol Sci 68:373–380.

Lichtenthaler HK (1996) Vegetation stress: an introduction to the stress concept in plants. J Plant Physiol 148:4–14. doi: 10.1016/S0176-1617(96)80287-2

Lichtenthaler HK (1998) The stress concept in plants: an introduction. Ann N Y Acad Sci 851:187–198. doi: 10.1111/j.1749-6632.1998.tb08993.x

Minina EA, Filonova LH, Daniel G, Bozhkov P V (2013) Detection and measurement of necrosis in plants. Methods Mol Biol 1004:229–248. doi: 10.1007/978-1-62703-383-1,1-17

Mittler R (2002) Oxidative stress, antioxidants and stress tolerance. Trends Plant Sci 7:405–410. doi: 10.1016/S1360-1385(02)02312-9

Munns R, James RA, Läuchli A (2006) Approaches to increasing the salt tolerance of wheat and other cereals. J Exp Bot 57:1025–1043. doi: 10.1093/jxb/erj100

Nah G, Pagliarulo CL, Mohr PG, et al (2009) Comparative sequence analysis of the SALT OVERLY SENSITIVE1 orthologous region in Thellungiella halophila and Arabidopsis thaliana. Genomics 94:196–203. doi: 10.1016/j.ygeno.2009.05.007

Noctor G, Foyer CH (1998) ASCORBATE AND GLUTATHIONE: Keeping Active Oxygen Under Control. Annu Rev Plant Physiol Plant Mol Biol 49:249–279. doi: 10.1146/annurev. arplant.49.1.249

Olien CR, Smith MN (1977) Ice adhesions in relation to freeze stress. Plant Physiol 60:499–503. doi: 10.1104/pp.60.4.499

Pahlich E (1993) Larcher's definition of plant stress: a valuable principle for metabolic adaptability research. Rev Bras Fisiol Veg 5:200–216.

Pearce RS (1999) Molecular analysis of acclimation to cold. Plant Growth Regul 29:47–76. doi: 10.1023/A:1006291330661

Pei L, Wang J, Li K, et al (2012) Overexpression of Thellungiella halophila H^+-pyrophosphatase gene improves low phosphate tolerance in maize. PLoS One 7:e43501. doi: 10.1371/journal. pone.0043501

Peterson JD, Umayam LA, Dickinson T, et al (2001) The comprehensive microbial resource. Nucleic Acids Res 29:123–125.

Peterson RKD, Higley LG (2000) Biotic Stress and Yield Loss, 1st edn. CRC Press, Boca Raton, FL

Reddy AR, Chaitanya KV, Vivekanandan M (2004) Drought-induced responses of photosynthesis and antioxidant metabolism in higher plants. J Plant Physiol 161:1189–1202. doi: 10.1016/j. jplph.2004.01.013

Rizhsky L, Davletova S, Liang H, Mittler R (2004) The zinc finger protein Zat12 is required for cytosolic ascorbate peroxidase 1 expression during oxidative stress in Arabidopsis. J Biol Chem 279:11736–11743. doi: 10.1074/jbc.M313350200

Rolland F, Baena-Gonzalez E, Sheen J (2006) Sugar sensing and signaling in plants: conserved and novel mechanisms. Annu Rev Plant Biol 57:675–709. doi: 10.1146/annurev. arplant.57.032905.105441

Saruyama N, Sakakura Y, Asano T, et al (2013) Quantification of metabolic activity of cultured plant cells by vital staining with fluorescein diacetate. Anal Biochem 441:58–62. doi: 10.1016/j.ab.2013.06.005

Sazzad K (2007) Exploring plant tolerance to biotic and abiotic stresses. Swedish University of Agricultural Sciences, Uppsala

Shabala S, Wu H, Bose J (2015) Salt stress sensing and early signalling events in plant roots: current knowledge and hypothesis. Plant Sci 241:109–119. doi: 10.1016/j.plantsci.2015.10.003

Shao H-B, Chu L-Y, Jaleel CA, et al (2009) Understanding water deficit stress-induced changes in the basic metabolism of higher plants—biotechnologically and sustainably improving agriculture and the ecoenvironment in arid regions of the globe. Crit Rev Biotechnol 29:131–151. doi: 10.1080/07388550902869792

Shao H-B, Guo Q-J, Chu L-Y, et al (2007) Understanding molecular mechanism of higher plant plasticity under abiotic stress. Colloids Surf B Biointerfaces 54:37–45. doi: 10.1016/j. colsurfb.2006.07.002

Shinozaki K, Yamaguchi-Shinozaki K (2000) Molecular responses to dehydration and low temperature: differences and cross-talk between two stress signaling pathways. Curr Opin Plant Biol 3:217–223.

Singh R, Jwa N-S (2013) Understanding the responses of rice to environmental stress using proteomics. J Proteome Res 12:4652–4669. doi: 10.1021/pr400689j

Somersalo S, Krause GH (1989) Photoinhibition at chilling temperature: fluorescence characteristics of unhardened and cold-acclimated spinach leaves. Planta 177:409–416. doi: 10.1007/ BF00403600

Song Y, Gao J, Yang F, et al (2013) Molecular evolutionary analysis of the Alfin-like protein family in Arabidopsis lyrata, Arabidopsis thaliana, and Thellungiella halophila. PLoS One 8:e66838. doi: 10.1371/journal.pone.0066838

Taiz L, Zeiger E (1991) Plant Physiology, 1st edn. Benjamin-Cummings Publishing Company Inc., San Francisco, CA

Taji T, Seki M, Satou M, et al (2004) Comparative genomics in salt tolerance between Arabidopsis and aRabidopsis-related halophyte salt cress using Arabidopsis microarray. Plant Physiol 135:1697–1709. doi: 10.1104/pp.104.039909

Tester M, Davenport R (2003) Na⁺ tolerance and Na⁺ Transport in higher plants. Ann Bot 91:503–527. doi: 10.1093/aob/mcg058

Thomashow MF (1998) Role of cold-responsive genes in plant freezing tolerance. Plant Physiol 118:1–8. doi: 10.1104/pp.118.1.1

Thomashow MF (1999) PLANT COLD ACCLIMATION: Freezing Tolerance Genes and Regulatory Mechanisms. Annu Rev Plant Physiol Plant Mol Biol 50:571–599. doi: 10.1146/annurev.arplant.50.1.571

Urao T (1999) A transmembrane hybrid-type histidine kinase in Arabidopsis functions as an osmosensor. Plant Cell 11:1743–1754. doi: 10.1105/tpc.11.9.1743

Vartapetian BB, Andreeva IN, Generozova IP, et al (2003) Functional electron microscopy in studies of plant response and adaptation to anaerobic stress. Ann Bot 91:155–172. doi: 10.1093/aob/mcf244

Verslues PE, Zhu J-K (2005) Before and beyond ABA: upstream sensing and internal signals that determine ABA accumulation and response under abiotic stress. Biochem Soc Trans 33:375–379. doi: 10.1042/BST0330375

Vialaret J, Di Pietro M, Hem S, et al (2014) Phosphorylation dynamics of membrane proteins from Arabidopsis roots submitted to salt stress. Proteomics 14:1058–1070. doi: 10.1002/pmic.201300443

Walling L (2000) The myriad plant responses to herbivores. J Plant Growth Regul 19:195–216.

Wang M, Wang Q, Zhang B (2013a) Evaluation and selection of reliable reference genes for gene expression under abiotic stress in cotton (Gossypium hirsutum L.). Gene 530:44–50. doi: 10.1016/j.gene.2013.07.084

Wang W, Wu Y, Li Y, et al (2010) A large insert Thellungiella halophila BIBAC library for genomics and identification of stress tolerance genes. Plant Mol Biol 72:91–99. doi: 10.1007/s11103-009-9553-3

Wang X, Chang L, Wang B, et al (2013b) Comparative proteomics of Thellungiella halophila leaves from plants subjected to salinity reveals the importance of chloroplastic starch and soluble sugars in halophyte salt tolerance. Mol Cell Proteomics 12:2174–2195. doi: 10.1074/mcp.M112.022475

Wardlaw IF (1972) Responses of plants to environmental stresses. J. Levitt. Academic Press, New York, 1972. xiv, 698 pp., illus. $32.50. Physiological Ecology. Science 177:786. doi:10.1126/science.177.4051.786

Zhu J-K (2001) Cell signaling under salt, water and cold stresses. Curr Opin Plant Biol 4(5):401–406.

Chapter 2
Omics and System Biology Approaches in Plant Stress Research

Abstract The continuous development of analytical and experimental technologies as well as instruments resulted in the development of very specialized experimental approaches that can identify, measure and quantify particular types of cellular molecules. These technologies are known as "Omics Technologies". Most of the omics technologies are high throughput with very fast data generation rates and humongous outputs. Thus, they are highly dependent on bioinformatics and computational tools. These technologies have made noticeable contributions to the current advancements in our understanding of plant biology in general and plant stress tolerance and response in particular. In this chapter, we will introduce the main omics technologies employed in plant biology and the bioinformatics platforms associated with them.

Keywords Plant stress • Biotic stress • Abiotic stress • Omics • Genomics • Proteomics • Proteogenomics • Transcriptomics • Metabolomics • Databases • Bioinformatics

2.1 Introduction

In the last two decades, molecular biology and systems biology experienced unprecedented advancements either in the accuracy of the analysis or at their overall scales (El-Metwally et al. 2014a). At first sight, molecular biology and systems biology look to be opposites due to the reductionistic nature of molecular biology and the holistic perspective of systems biology respectively. However, in modern research settings, both molecular and systems biology complement each other, providing new perspectives in approaching complex topics such as the study of plant stress or the improvement of plant stress responses (Duque et al. 2013). Figure 2.1 shows the common approaches utilized in plant stress research including omics-based researches.

Genome sequencing represented a landmark in the development of biological sciences and the methodologies of approaching biological problems. It enhanced our grasp of biological systems by examining the base of life (i.e. DNA) and allowed an enriched understanding of gene structures and functions (El-Metwally et al. 2014a). Next-generation sequencing (NGS) of the genome elevates the utility of genome sequencing by providing cheap, fast and easy genome sequencing platforms, though not without some challenges (El-Metwally et al. 2013, 2014a). NGS presents a great

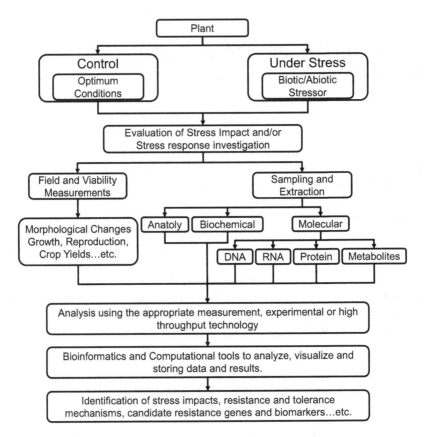

Fig. 2.1 Schematic overview of common approaches in plant stress research

foundation for several other methodologies to be developed as well as several new approaches for studying biological systems in the so-called post-genomic era (Duque et al. 2013).

Genomics, transcriptomics, proteomics, proteogenomics and metabolomics are modern methodologies and approaches that have been recently applied in the study of plant stress mechanism responses. They provide new insights and open new horizons for understanding stresses and responses as well as the improvement of plant responses and resistance to stresses (Duque et al. 2013). Due to the large-scale nature of these approaches, bioinformatics and computational approaches are highly associated with the above for either developing new data analytical methods, better visualization or storage in sustainable online resources (Helmy et al. 2011, 2012a, b, c). Figure 2.2 shows the main omics approaches employed in plant stress research, its primary technologies and expected outcomes.

Since the focus of this book is the integrated omics approaches in plant stress tolerance, we will introduce the applications of each of the omics and bioinformatics approaches in detail.

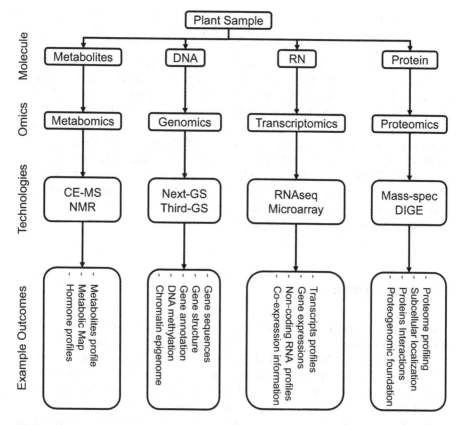

Fig. 2.2 Schematic overview of the main molecular and systems biology approaches and their technologies as well as expected outcomes in plant biology and stress research

2.2 Genomics

Genomics is the study of all the genes in a given genome including the identification of gene sequences, intragenic sequences, gene structures and annotations (Duque et al. 2013). The technology of choice for genomics is genome sequencing that began with the first generation of methods in the 1970s, followed by the next-generation sequencing (NGS) technologies in the middle 1990s as well as the more recent third-generation sequencing technologies (El-Metwally et al. 2014b, c). The process involves DNA extraction, amplification using polymerase chain reaction (PCR) techniques, DNA sequencing and sequence assembly as well as quality assessment (El-Metwally et al. 2013, 2014a, d). Following DNA sequencing and assembly, the gene structural and functional annotation takes place revealing invaluable information about the biology of the organism in question.

It would be challenging to list all the contributions of genomics to the study of the plant stress response and tolerance research. Therefore, we will list the main applications of genomics in this field. Genomics mainly helped in identifying the functional relevance of genes involved in abiotic and biotic stress responses in plants via functional genomic approaches (Cao et al. 2005; Govind et al. 2009; Ramegowda et al. 2013, 2014). Combined with other techniques, genomics helped plant breeders create new breeds that can tolerate several biotic and abiotic stresses and, consequently, have increased crop yields. This includes new breeds for drought and cold tolerance as well as pathogen resistance (Yadav et al. 2010; Yao et al. 2011; Le et al. 2012; Chen et al. 2012; Shankar et al. 2013; Wang et al. 2013; Agarwal et al. 2014). Furthermore, the huge online databases, repositories and archives of plant genomic information serve as a foundation for transcriptomics, genome engineering and proteogenomics (Matthews et al. 2009; Batley and Edwards 2009; Mochida and Shinozaki 2010; Jung and Main 2014).

2.3   Transcriptomics

The transcriptome is the RNA expression profile of an organism. Unlike the genome which remains constant despite age, organ or growth conditions, the transcriptome is highly dynamic (El-Metwally et al. 2014a). Therefore, the term transcriptomics refers to the capturing of the RNA expression profile in spatial and temporal bases in certain plant organs, tissues and cells within particular context (Duque et al. 2013; El-Metwally et al. 2014a). This particular context can be growth or environmental conditions, treatment with certain nutrients or biotic stress conditions. The RNA/gene expression profiling is mostly accomplished using microarray, RNA sequencing (RNAseq) through next-generation sequencing (NGS), serial analysis of gene expression (SAGE) and digital gene expression profiling (Kawahara et al. 2012; Duque et al. 2013; De Cremer et al. 2013).

Transcriptomics mainly helps in finding genes that are associated with alterations in the plant phenotype under different conditions. For instance, transcriptomics can be used in finding candidate genes that contribute to stress tolerance through the comparison of transcriptomes of the same plant under optimal and stress conditions (Le et al. 2012; Zhang et al. 2014b). Transcriptomics applications in plant stress response and tolerance can also include searching for abiotic stress candidate genes, predicting tentative gene functions and providing a better understanding of the plant-pathogen relationship (Kawahara et al. 2012; Jogaiah et al. 2013; De Cremer et al. 2013; Agarwal et al. 2014). The recent increase in the availability of online resources, databases and archives of transcriptome data allows for performing novel genome-wide analysis of plant stress responses and tolerances (Mochida and Shinozaki 2011; Duque et al. 2013; Jogaiah et al. 2013).

2.4 Proteomics

The proteome is the total expressed protein under certain conditions in a given organism, organ, cell, tissue or microorganism population (Tyers and Mann 2003). Similar to the transcriptome, the proteome is highly dynamic and changes based on temporal and environmental factors. Proteomics are the techniques used in capturing and measuring (or "profiling") the expressed proteins in a specific context (Tyers and Mann 2003). There are several types of proteomes that can be measured, and each of them can reveal particular information about the expressed proteins. The most common proteomes to be measured in plant stress tolerance and other plant related studies are the whole proteome and the phosphoproteome (Helmy et al. 2011, 2012b, c). The whole proteome is the quantitative and/or qualitative profiling of all the expressed proteins in a given sample, while the phosphoproteome is the quantitative and/or qualitative profiling of the phosphorylated proteins expressed in a given sample (Nakagami et al. 2012).

The technology of choice for proteomics is mass spectrometry (MS) including several approaches such as liquid chromatography–mass spectrometry (LC-MS/MS), Ion Trap–mass spectrometry (IT-MS) and matrix-assisted laser desorption/ionization–mass spectrometry (MLDI-MS) (Helmy et al. 2011, 2012a; Komatsu et al. 2014; Shao et al. 2014). These technologies are basically used in measuring the mass and charge of small protein fragments (or "peptides") that result from protein enzymatic digestion with special enzymes called proteases, such as trypsin (Helmy et al. 2011; Nakagami et al. 2012). The output of a standard MS-based proteomic analysis is a set of peptide fingerprints called MS spectra. MS spectra require another layer of interpretation to reveal the peptide sequences associated with each of them, the protein of each peptide and the modification occurring in each protein after being translated (Tyers and Mann 2003; Helmy et al. 2012c; Nakagami et al. 2012). Furthermore, several proteomics labs use protein electrophoresis technologies such as two-dimensional electrophoresis and Difference Gel Electrophoresis (DIGE) in plant proteomics (Cramer and Westermeier 2012; Duque et al. 2013; Komatsu et al. 2014; Arentz et al. 2014).

Proteomics is a very informative approach that is used to reveal invaluable information when studying plant stress response and tolerance, either in a genome-wide or sample scale (Nakagami et al. 2012). It can be used to profile all the expressed proteins in multiple conditions (e.g. optimal, stress and prolonged stress conditions) and cross compare these different sets to pinpoint the proteins involved in stress tolerance (Evers et al. 2012; Yan et al. 2014). Quantitative proteomics reveals the proteins that are differentially expressed under the condition changes, which points towards its contribution in the stress response process as well (Liu et al. 2015). Phosphoproteomics is more associated with the identification of proteins activated and functioning under certain condition. Therefore, it is very useful in identifying pathways involved in a particular function or process through ascertaining the set of proteins that are exclusively activated under the condition that triggered this function. Through phosphoproteomics, proteins and signaling pathways involved in response to particular stress can be identified (Sugiyama et al. 2008; Lassowskat et al. 2014; Zhang et al. 2014a). Both whole proteomics and phosphoproteomics can be combined in one comprehensive study to provide a better understanding of the stress in question (Margaria et al. 2013; Yang et al. 2013; Hopff et al. 2013).

2.5 Proteogenomics

Proteogenomics is a comprehensive approach that combines large-scale proteomic data with genomic and/or transcriptomics data in genome annotation refinement and the elucidation of novel regulatory mechanisms (Helmy et al. 2012a; Ansong et al. 2008). The proteomics data generated by means of MS-based proteomics is well known for its high throughput and accuracy. Therefore, it provides a rich source of translation-level information about the expressed proteins and can be used as a source of affordable large-scale experimental evidence for several predictions (Helmy et al. 2012b, c; Ansong et al. 2008; de Groot et al. 2009; Armengaud 2010). In a standard proteogenomics study, the naturally expressed proteins are identified using MS-based proteogenomics followed by mapping them back to the genomic or transcriptomic data (Helmy et al. 2012a; Ansong et al. 2008).

In the last decade, proteogenomics has helped in elevating our understanding of the biology of plants in general as well as plant stress research in particular. For instance, a large-scale proteogenomics study of *Arabidopsis thaliana* identified 57 new genes and corrected the annotations of hundreds of its genes using intensive sampling from the *Arabidopsis* organs under several conditions and in different life stages (Baerenfaller et al. 2008). Another study reported corrections and new identifications in about 13% of the annotated genes in *Arabidopsis* (Castellana et al. 2008). Furthermore, several major cultivated crops such as *Oryza sativa* and *Zea mays* benefited from proteogenomics studies (Helmy et al. 2011; Castellana et al. 2014).

In plant stress research, the use of proteogenomics provided a deeper understanding of the major abiotic stress factors including bacteria such as *Bradyrhizobium diazoefficiens* (Chapman and Bellgard 2014), fungi such as *Aspergillus niger* (the black mold fungus) and *Stagonospora nodorum* (Wright et al. 2009; Bringans et al. 2009), insects such as *Drosophila melanogaster* (Tress et al. 2008; Loevenich et al. 2009) and nematodes such as *Pristionchus pacificus* (Borchert et al. 2010). It also presented new insight into the investigation of the host-pathogen relationship such as understanding the relationship between the plant and the phyllosphere bacteria (Delmotte et al. 2009), identifying novel effectors in fungal diseases (Cooke et al. 2014), providing a better understanding of the host-parasite relationship (Lasonder et al. 2002; Bindschedler et al. 2009) as well as shedding light on the mechanisms of environmental adaptation and ecological diversity (de Groot et al. 2009; Denef et al. 2010).

2.6 Metabolomics

The metabolome is the complete set of metabolites that can be identified in a given organism, organ, tissue or biological fluid. Thus, metabolomics refers to techniques and methods used to study the metabolome (Duque et al. 2013). Due to differences in the chemical and physical properties of the metabolites, a combination of several analytical and separation techniques is required to obtain the metabolic profile of a plant or given sample (Jogaiah et al. 2013). Although Capillary Electrophoresis-liquid-chromatography Mass Spectrometry (CE-MS) is considered the most advanced metabolomics technology to date (Soga et al. 2002), several other analytical instruments and separation technologies are employed in metabolomics such as Gas Chromatography (GC), Mass Spectrometry (MS) and Nuclear Magnetic Resonance (NMR) (Saito and Matsuda 2010; Duque et al. 2013; Jogaiah et al. 2013).

Plants are able to synthesize a wide spectrum of chemical and biological compounds that are crucial for regulating the response to different types of biotic and abiotic stress. Therefore, identifying the metabolites produced by the plant under each stress condition provides a better understanding of the regulation processes as well as joins the genotype with the phenotype and investigates the changes in phenotype that take place under stress conditions (Badjakov et al. 2012). Metabolomics is usually used in combination with other omics analysis (e.g. transcriptomics or proteomics) to investigate the correlation between metabolite levels and the expression level of genes/proteins (Srivastava et al. 2013). A strong correlation between stress metabolites and a certain gene/protein indicates the role of this gene/protein in the response process (Urano et al. 2010; Duque et al. 2013; Jogaiah et al. 2013). Metabolomics is used to provide a better understanding of the stress response and tolerance process in model plants such as *Arabidopsis* (Cook et al. 2004) as well as production crops such as the common bean (*Phaseolus vulgaris*) (Broughton et al. 2003), poplars (*Populus x canescens*) (Behnke et al. 2010), cereals (Sicher and Barnaby 2012) and other food crops (Hernández et al. 2007; Duque et al. 2013).

2.7 Bioinformatics

The brief introduction of each of the omics approaches that we provided above shows that all of them share similar high throughput and large-scale properties. Furthermore, these approaches can be genome-wide as well as through the combination of several genomes or several species, which results in producing huge amounts of data that requires proper handling, analysis, visualization and storage (El-Metwally et al. 2014e). Therefore, all omics research is tightly bound with strong bioinformatics and computational tools that perform the various analysis tasks as well as allow integration between several types of data "multi-omics" and enable knowledge exchange between different organisms (Shinozaki and Sakakibara 2009; Mochida and Shinozaki 2011; El-Metwally et al. 2014a).

2.7.1   Data Handling and Analysis

The primary reason for including informatics analysis and computational tools as well as associated methods and algorithms in biology is to allow biological data analysis in an accurate, fast, human-error free and easily reproducible manner (Orozco et al. 2013). Hence, several bioinformatics tasks became indispensable in biological research in general and plant stress multi-omics research in particular. This includes the standard tasks involved in genome sequence assembly (El-Metwally et al. 2013, 2014f, g), sequence alignment (Altschul et al. 1990; Tatusova and Madden 1999), gene prediction (Stanke and Morgenstern 2005), peptides and pertains sequence identification (MS-spectra interpretation tools) (Eng et al. 1994; Perkins et al. 1999), gene and protein function prediction (Falda et al. 2012; Yachdav et al. 2014), DNA-protein and protein-protein interaction prediction (McDowall et al. 2009; Franceschini et al. 2013), interaction and regulatory networks analysis (Chaouiya 2012) and several other essential tasks (Polpitiya et al. 2008; Henry et al. 2014).

2.7.2   Data and Results Visualization

The large amount of data generated by modern analytical and experimental instruments such as genome sequencers and mass spectrometers as well as the information resulting from the analysis and processing of this data requires special types of visualization. Thus, several tools were developed to help visualize the biological data and results in a manner that would maximize the utility of the data. These genomic data visualizing tools include Gbrowse, UCSC Genome Browser and Integrated Genome Viewer (IGV) (Stein et al. 2002; Karolchik et al. 2003; Pang et al. 2014), proteomics data visualization tools such as PRIDE Inspector and ConPath (Kim et al. 2008; Wang et al. 2012), proteogenomics data or multi-omics data visualization tools such as PGFeval, 3Omics, Peppy (Helmy et al. 2011, 2012b; Kuo et al. 2013; Risk et al. 2013), metabolomics visualization tools such as MultiExperiment Viewer (MeV) (Saeed et al. 2003) and network visualization tools such as Cytoscape and its associated web versions Cytoscape.js and Cytoscape web (Shannon et al. 2003; Lopes et al. 2010; Ono et al. 2014).

2.7.3   Data and Results Storage and Maintenance

High throughput data is very fruitful in that we can gain more knowledge from it by applying different types of analyses or by combining several datasets into one large-scale comparative analysis. However, this requires the data and results to be sustainably available and accessible to the scientific community (Smalter Hall et al. 2013; Helmy et al. 2016). Therefore, several types of databases are available online for depositing and storing the biological data and results. The databases range from those that store plant information as classification, growth, production, geographical distribution (Wilkinson et al. 2012), plant genomic information (Yu et al. 2013; Zhao et al. 2014), plant transcriptomic information (Priya and Jain 2013), plant proteomic information (Komatsu and Tanaka 2005; Cheng et al. 2014), plant proteogenomic information (Helmy et al. 2011, 2012b) and plant metabolomic information (Deborde and Jacob 2014). Furthermore, some databases are specialized in storing and maintaining plant stress resistance and tolerance information such as STIFDB2, the Arabidopsis stress responsive gene database, QlicRice and the fungal stress response database (FSRD) (Smita et al. 2011; Borkotoky et al. 2013; Karányi et al. 2013; Naika et al. 2013).

In general, several tools provide more than one of the above-mentioned features such as data analysis and visualization (Eng et al. 1994; Perkins et al. 1999; Cargile et al. 2004; Helmy et al. 2011) or data visualization and storage (Helmy et al. 2012b). Furthermore, several tools provide these services for multi-omics data and results (Kuo et al. 2013).

References

Agarwal P, Parida SK, Mahto A, et al (2014) Expanding frontiers in plant transcriptomics in aid of functional genomics and molecular breeding. Biotechnol J 9:1480–1492. doi:10.1002/biot.201400063

Altschul SF, Gish W, Miller W, et al (1990) Basic local alignment search tool. J Mol Biol 215:403–410.

Ansong C, Purvine SO, Adkins JN, et al (2008) Proteogenomics: needs and roles to be filled by proteomics in genome annotation. Br Funct Genomic Proteomic 7:50–62.

Arentz G, Weiland F, Oehler MK, Hoffmann P (2014) State of the art of 2D DIGE. Proteomics Clin Appl 9:277-288. doi:10.1002/prca.201400119

Armengaud J (2010) Proteogenomics and systems biology: quest for the ultimate missing parts. Expert Rev Proteomics 7:65–77. doi:10.1586/epr.09.104

Baerenfaller K, Grossmann J, Grobei MA, et al (2008) Genome-scale proteomics reveals *Arabidopsis thaliana* gene models and proteome dynamics. Science 320:938–941

Batley J, Edwards D (2009) Genome sequence data: management, storage, and visualization. Biotechniques 46:333–334., 336. doi:10.2144/000113134

Behnke K, Kaiser A, Zimmer I, et al (2010) RNAi-mediated suppression of isoprene emission in poplar transiently impacts phenolic metabolism under high temperature and high light intensities: a transcriptomic and metabolomic analysis. Plant Mol Biol 74:61–75. doi:10.1007/s11103-010-9654-z

Bindschedler L V, Burgis TA, Mills DJS, et al (2009) In planta proteomics and proteogenomics of the biotrophic barley fungal pathogen Blumeria graminis f. sp. hordei. Mol Cell Proteomics MCP 8:2368–2381. doi:10.1074/mcp.M900188-MCP200

Borchert N, Dieterich C, Krug K, et al (2010) Proteogenomics of Pristionchus pacificus reveals distinct proteome structure of nematode models. Genome Res 20:837–846. doi:10.1101/gr.103119.109

Borkotoky S, Saravanan V, Jaiswal A, et al (2013) The Arabidopsis stress responsive gene database. Int J Plant Genomics 2013:949564. doi:10.1155/2013/949564

Bringans S, Hane JK, Casey T, et al (2009) Deep proteogenomics; high throughput gene validation by multidimensional liquid chromatography and mass spectrometry of proteins from the fungal wheat pathogen Stagonospora nodorum. BMC Bioinformatics 10:301. doi:10.1186/1471-2105-10-301

Broughton WJ, Hernández G, Blair M, et al (2003) Beans (Phaseolus spp.)—model food legumes. Plant and Soil 252:55–128. doi:10.1023/A:1024146710611

Cao X, Zhou P, Zhang X, et al (2005) Identification of an RNA silencing suppressor from a plant double-stranded RNA virus. J Virol 79:13018–13027. doi:10.1128/JVI.79.20.13018-13027.2005

Cargile BJ, Bundy JL, Freeman TW, Stephenson Jr. JL (2004) Gel based isoelectric focusing of peptides and the utility of isoelectric point in protein identification. J Proteome Res 3:112–119.

Castellana NE, Payne SH, Shen Z, et al (2008) Discovery and revision of Arabidopsis genes by proteogenomics. Proc Natl Acad Sci U S A 105:21034–21038.

Castellana NE, Shen Z, He Y, et al (2014) An automated proteogenomic method uses mass spectrometry to reveal novel genes in Zea mays. Mol Cell Proteomics MCP 13:157–167. doi:10.1074/mcp.M113.031260

Chapman B, Bellgard M (2014) High-throughput parallel proteogenomics: a bacterial case study. Proteomics 14:2780–2789. doi:10.1002/pmic.201400185

Chen S, Jiang J, Li H, Liu G (2012) The salt-responsive transcriptome of Populus simonii × Populus nigra via DGE. Gene 504:203–212. doi:10.1016/j.gene.2012.05.023

Cheng H, Deng W, Wang Y, et al (2014) dbPPT: a comprehensive database of protein phosphorylation in plants. Database 2014:bau121. doi:10.1093/database/bau121

Claudine Chaouiya (2012). Logical Modelling of Gene Regulatory Networks with GINsim in Methods in molecular biology Edited by N.J. Clifton, Humana Press, Print ISBN 1940-6029

Cook D, Fowler S, Fiehn O, Thomashow MF (2004) A prominent role for the CBF cold response pathway in configuring the low-temperature metabolome of Arabidopsis. Proc Natl Acad Sci U S A 101:15243–15248. doi:10.1073/pnas.0406069101

Cooke IR, Jones D, Bowen JK, et al (2014) Proteogenomic analysis of the Venturia pirina (Pear Scab Fungus) secretome reveals potential effectors. J Proteome Res 13:3635–3644. doi:10.1021/pr500176c

Cramer, Rainer, Westermeier R (2012) Difference Gel Electrophoresis (DIGE) - Methods and Protocols. Humana Press, Print ISBN: 9781617795732

De Cremer K, Mathys J, Vos C, et al (2013) RNAseq-based transcriptome analysis of Lactuca sativa infected by the fungal necrotroph Botrytis cinerea. Plant Cell Environ 36:1992–2007. doi:10.1111/pce.12106

de Groot A, Dulermo R, Ortet P, et al (2009) Alliance of proteomics and genomics to unravel the specificities of Sahara bacterium Deinococcus deserti. PLoS Genet 5:e1000434. doi:10.1371/journal.pgen.1000434

Deborde C, Jacob D (2014) MeRy-B, a metabolomic database and knowledge base for exploring plant primary metabolism. Methods Mol Biol 1083:3–16. doi:10.1007/978-1-62703-661-0_1

Delmotte N, Knief C, Chaffron S, et al (2009) Community proteogenomics reveals insights into the physiology of phyllosphere bacteria. Proc Natl Acad Sci U S A 106:16428–16433. doi:10.1073/pnas.0905240106

Denef VJ, Kalnejais LH, Mueller RS, et al (2010) Proteogenomic basis for ecological divergence of closely related bacteria in natural acidophilic microbial communities. Proc Natl Acad Sci U S A 107:2383–2390. doi:10.1073/pnas.0907041107

Duque AS, de Almeida AM, da Silva AB, da Silva JM, et al (2013) Abiotic stress—plant responses and applications in agriculture. doi:10.5772/45842

El-Metwally S, Hamza T, Zakaria M, Helmy M (2013) Next-generation sequence assembly: four stages of data processing and computational challenges. PLoS Comput Biol 9:e1003345. doi:10.1371/journal.pcbi.1003345

El-Metwally S, Ouda OM, Helmy M (2014a) Next generation sequencing technologies and challenges in sequence assembly.

El-Metwally S, Ouda OM, Helmy M (2014b) First- and next-generations sequencing methods. Next Gener Seq Technol Challenges Seq Assem. doi:10.1007/978-1-4939-0715-1_3

El-Metwally S, Ouda OM, Helmy M (2014c) New horizons in next-generation sequencing. Next Gener Seq Technol Challenges Seq Assem. doi:10.1007/978-1-4939-0715-1_6

El-Metwally S, Ouda OM, Helmy M (2014d) Assessment of next-generation sequence assembly. Next Gener Seq Technol Challenges Seq Assem. doi:10.1007/978-1-4939-0715-1_10

El-Metwally S, Ouda OM, Helmy M (2014e) Novel next-generation sequencing applications. Next Gener Seq Technol Challenges Seq Assem. doi:10.1007/978-1-4939-0715-1_7

El-Metwally S, Ouda OM, Helmy M (2014f) Next-generation sequence assembly overview. Next Gener Seq Technol Challenges Seq Assem. doi:10.1007/978-1-4939-0715-1_8

El-Metwally S, Ouda OM, Helmy M (2014g) Next-generation sequence assemblers. Next Gener Seq Technol Challenges Seq Assem. doi:10.1007/978-1-4939-0715-1_11

Eng JK, McCormack AL, Yates JR (1994) An approach to correlate tandem mass spectral data of peptides with amino acid sequences in a protein database. J Am Soc Mass Spectrom 5:976–989. doi:10.1016/1044-0305(94)80016-2

Evers D, Legay S, Lamoureux D, et al (2012) Towards a synthetic view of potato cold and salt stress response by transcriptomic and proteomic analyses. Plant Mol Biol 78:503–514. doi:10.1007/s11103-012-9879-0

Falda M, Toppo S, Pescarolo A, et al (2012) Argot2: a large scale function prediction tool relying on semantic similarity of weighted Gene Ontology terms. BMC Bioinformatics 13(Suppl 4):S14. doi:10.1186/1471-2105-13-S4-S14

Franceschini A, Szklarczyk D, Frankild S, et al (2013) STRING v9.1: protein-protein interaction networks, with increased coverage and integration. Nucleic Acids Res 41:D808–15. doi:10.1093/nar/gks1094

Govind G, Harshavardhan VT, Patricia JK, et al (2009) Identification and functional validation of a unique set of drought induced genes preferentially expressed in response to gradual water stress in peanut. Mol Genet Genomics 281:607. doi:10.1007/s00438-009-0441-y

Helmy M, Tomita M, Ishihama Y (2011) OryzaPG-DB: rice proteome database based on shotgun proteogenomics. BMC Plant Biol 11:63. doi:10.1186/1471-2229-11-63

Helmy M, Sugiyama N, Tomita M, Ishihama Y (2012a) Mass spectrum sequential subtraction speeds up searching large peptide MS/MS spectra datasets against large nucleotide databases for proteogenomics. Cell Mech 17:633–644. doi:10.1111/j.1365-2443.2012.01615.x

Helmy M, Sugiyama N, Tomita M, Ishihama Y (2012b) The rice proteogenomics database oryza PG-DB: development, expansion, and new features. Front Plant Sci 3:65. doi:10.3389/fpls.2012.00065

Helmy M, Tomita M, Ishihama Y (2012c) Peptide identification by searching large-scale tandem mass spectra against large databases: bioinformatics methods in proteogenomics. Gene Genome Genomics 6:76–85.

Helmy M, Crits-Christoph A, Bader GD, et al (2016) Ten simple rules for developing public biological databases. PLoS Comput Biol 12:e1005128. doi:10.1371/journal.pcbi.1005128

Henry VJ, Bandrowski AE, Pepin A-S, et al (2014) OMICtools: an informative directory for multi-omic data analysis. Database (Oxford) 2014:bau069. doi:10.1093/database/bau069

Hernández G, Ramírez M, Valdés-López O, et al (2007) Phosphorus stress in common bean: root transcript and metabolic responses. Plant Physiol 144:752–767. doi:10.1104/pp.107.096958

Hopff D, Wienkoop S, Lüthje S (2013) The plasma membrane proteome of maize roots grown under low and high iron conditions. J Proteomics 91:605–618. doi:10.1016/j.jprot.2013.01.006

Ilian Badjakov, Violeta Kondakova and Atanas Atanassov (2012). Current View on Fruit Quality in Relation to Human Health in Sustainable Agriculture and New Biotechnologies, Edited by Noureddine Benkeblia, CRC Press, Boca Raton, Pages 303–319, Print ISBN: 978-1-4398-2504-4, eBook ISBN: 978-1-4398-2505-1. doi: 10.1201/b10977-14

Jogaiah S, Govind SR, Tran L-SP (2013) Systems biology-based approaches toward understanding drought tolerance in food crops. Crit Rev Biotechnol 33:23–39. doi:10.3109/07388551.2012.659174

Jung S, Main D (2014) Genomics and bioinformatics resources for translational science in Rosaceae. Plant Biotechnol Rep 8:49–64. doi:10.1007/s11816-013-0282-3

Karányi Z, Holb I, Hornok L, et al (2013) FSRD: fungal stress response database. Database (Oxford) 2013:bat037. doi:10.1093/database/bat037

Karolchik D, Baertsch R, Diekhans M, et al (2003) The UCSC genome browser database. Nucleic Acids Res 31:51–54.

Kawahara Y, Oono Y, Kanamori H, et al (2012) Simultaneous RNA-seq analysis of a mixed transcriptome of rice and blast fungus interaction. PLoS One 7:e49423. doi:10.1371/journal.pone.0049423

Kim P-G, Cho H-G, Park K (2008) A scaffold analysis tool using mate-pair information in genome sequencing. J Biomed Biotechnol 2008:675741. doi:10.1155/2008/675741

Komatsu S, Tanaka N (2005) Rice proteome analysis: a step toward functional analysis of the rice genome. Proteomics 5:938–949.

Komatsu S, Kamal AHM, Hossain Z (2014) Wheat proteomics: proteome modulation and abiotic stress acclimation. Front Plant Sci 5:684. doi:10.3389/fpls.2014.00684

Kuo T-C, Tian T-F, Tseng YJ (2013) 3Omics: a web-based systems biology tool for analysis, integration and visualization of human transcriptomic, proteomic and metabolomic data. BMC Syst Biol 7:64. doi:10.1186/1752-0509-7-64

Lasonder E, Ishihama Y, Andersen JS, et al (2002) Analysis of the Plasmodium falciparum proteome by high-accuracy mass spectrometry. Nature 419:537–542.

Lassowskat I, Böttcher C, Eschen-Lippold L, et al (2014) Sustained mitogen-activated protein kinase activation reprograms defense metabolism and phosphoprotein profile in Arabidopsis thaliana. Front Plant Sci 5:554. doi:10.3389/fpls.2014.00554

Le DT, Nishiyama R, Watanabe Y, et al (2012) Differential gene expression in soybean leaf tissues at late developmental stages under drought stress revealed by genome-wide transcriptome analysis. PLoS One 7:e49522. doi:10.1371/journal.pone.0049522

Liu B, Zhang N, Zhao S, et al (2015) Proteomic changes during tuber dormancy release process revealed by iTRAQ quantitative proteomics in potato. Plant Physiol Biochem 86:181–190. doi:10.1016/j.plaphy.2014.12.003

Loevenich SN, Brunner E, King NL, et al (2009) The Drosophila melanogaster PeptideAtlas facilitates the use of peptide data for improved fly proteomics and genome annotation. BMC Bioinformatics 10:59. doi:10.1186/1471-2105-10-59

Lopes CT, Franz M, Kazi F, et al (2010) Cytoscape web: an interactive web-based network browser. Bioinformatics 26:2347–2348. doi:10.1093/bioinformatics/btq430

Margaria P, Abbà S, Palmano S (2013) Novel aspects of grapevine response to phytoplasma infection investigated by a proteomic and phospho-proteomic approach with data integration into functional networks. BMC Genomics 14:38. doi:10.1186/1471-2164-14-38

Matthews DE, Lazo GR, Anderson OD (2009) Plant and crop databases. Methods Mol Biol 513:243–262. doi:10.1007/978-1-59745-427-8_13

McDowall MD, Scott MS, Barton GJ (2009) PIPs: human protein-protein interaction prediction database. Nucleic Acids Res 37:D651–6. doi:10.1093/nar/gkn870

Mochida K, Shinozaki K (2010) Genomics and bioinformatics resources for crop improvement. Plant Cell Physiol 51:497–523. doi:10.1093/pcp/pcq027

Mochida K, Shinozaki K (2011) Advances in omics and bioinformatics tools for systems analyses of plant functions. Plant Cell Physiol 52:2017–2038. doi:10.1093/pcp/pcr153

Naika M, Shameer K, Mathew OK, et al (2013) STIFDB2: an updated version of plant stress-responsive transcription factor database with additional stress signals, stress-responsive transcription factor binding sites and stress-responsive genes in Arabidopsis and rice. Plant Cell Physiol 54:e8. doi:10.1093/pcp/pcs185

Nakagami H, Sugiyama N, Ishihama Y, Shirasu K (2012) Shotguns in the front line: phosphoproteomics in plants. Plant Cell Physiol 53:118–124. doi:10.1093/pcp/pcr148

Ono K, Demchak B, Ideker T (2014) Cytoscape tools for the web age: D3.js and cytoscape.js exporters. F1000Research 3:143. doi:10.12688/f1000research.4510.2

Orozco A, Morera J, Jiménez S, Boza R (2013) A review of bioinformatics training applied to research in molecular medicine, agriculture and biodiversity in Costa Rica and Central America. Brief Bioinform 14:661–670. doi:10.1093/bib/bbt033

Pang CNI, Tay AP, Aya C, et al (2014) Tools to covisualize and coanalyze proteomic data with genomes and transcriptomes: validation of genes and alternative mRNA splicing. J Proteome Res 13:84–98. doi:10.1021/pr400820p

Perkins DN, Pappin DJ, Creasy DM, Cottrell JS (1999) Probability-based protein identification by searching sequence databases using mass spectrometry data. Electrophoresis 20:3551–3567.

Polpitiya AD, Qian W-J, Jaitly N, et al (2008) DAnTE: a statistical tool for quantitative analysis of -omics data. Bioinformatics 24:1556–1558. doi:10.1093/bioinformatics/btn217

Priya P, Jain M (2013) RiceSRTFDB: a database of rice transcription factors containing comprehensive expression, cis-regulatory element and mutant information to facilitate gene function analysis. Database (Oxford) 2013:bat027. doi:10.1093/database/bat027

Ramegowda V, Senthil-kumar M, Udayakumar M, Mysore KS (2013) A high-throughput virus-induced gene silencing protocol identifies genes involved in multi-stress tolerance. BMC Plant Biol 13:193. doi:10.1186/1471-2229-13-193

Ramegowda V, Mysore KS, Senthil-Kumar M (2014) Virus-induced gene silencing is a versatile tool for unraveling the functional relevance of multiple abiotic-stress-responsive genes in crop plants. Front Plant Sci 5:323. doi:10.3389/fpls.2014.00323

Risk BA, Spitzer WJ, Giddings MC (2013) Peppy: proteogenomic search software. J Proteome Res 12:3019–3025. doi:10.1021/pr400208w

Saeed AI, Sharov V, White J, et al (2003) TM4: a free, open-source system for microarray data management and analysis. Biotechniques 34:374–378.

Saito K, Matsuda F (2010) Metabolomics for functional genomics, systems biology, and biotechnology. Annu Rev Plant Biol 61:463–489. doi:10.1146/annurev.arplant.043008.092035

Shankar A, Singh A, Kanwar P, et al (2013) Gene expression analysis of rice seedling under potassium deprivation reveals major changes in metabolism and signaling components. PLoS One 8:e70321. doi:10.1371/journal.pone.0070321

Shannon P, Markiel A, Ozier O, et al (2003) Cytoscape: a software environment for integrated models of biomolecular interaction networks. Genome Res 13:2498–2504. doi:10.1101/gr.1239303

Shao S, Guo T, Aebersold R (2014) Mass spectrometry-based proteomic quest for diabetes biomarkers. Biochim Biophys Acta doi:10.1016/j.bbapap.2014.12.012

Shinozaki K, Sakakibara H (2009) Omics and bioinformatics: an essential toolbox for systems analyses of plant functions beyond 2010. Plant Cell Physiol 50:1177–1180. doi:10.1093/pcp/pcp085

Sicher RC, Barnaby JY (2012) Impact of carbon dioxide enrichment on the responses of maize leaf transcripts and metabolites to water stress. Physiol Plant 144:238–253. doi:10.1111/j.1399-3054.2011.01555.x

Smalter Hall A, Shan Y, Lushington G, Visvanathan M (2013) An overview of computational life science databases & exchange formats of relevance to chemical biology research. Comb Chem High Throughput Screen 16:189–198

Smita S, Lenka SK, Katiyar A, et al (2011) QlicRice: a web interface for abiotic stress responsive QTL and loci interaction channels in rice. Database (Oxford) 2011:bar037. doi:10.1093/database/bar037

Soga T, Ueno Y, Naraoka H, et al (2002) Simultaneous determination of anionic intermediates for Bacillus subtilis metabolic pathways by capillary electrophoresis electrospray ionization mass spectrometry. Anal Chem 74:2233–2239.

Srivastava V, Obudulu O, Bygdell J, et al (2013) OnPLS integration of transcriptomic, proteomic and metabolomic data shows multi-level oxidative stress responses in the cambium of transgenic hipI- superoxide dismutase Populus plants. BMC Genomics 14:893. doi:10.1186/1471-2164-14-893

Stanke M, Morgenstern B (2005) AUGUSTUS: a web server for gene prediction in eukaryotes that allows user-defined constraints. Nucleic Acids Res 33:W465–7. doi:10.1093/nar/gki458

Stein LD, Mungall C, Shu S, et al (2002) The generic genome browser: a building block for a model organism system database. Genome Res 12:1599–1610.

Sugiyama N, Nakagami H, Mochida K, et al (2008) Large-scale phosphorylation mapping reveals the extent of tyrosine phosphorylation in Arabidopsis. Mol Syst Biol 4:193.

Tatusova TA, Madden TL (1999) BLAST 2 Sequences, a new tool for comparing protein and nucleotide sequences. FEMS Microbiol Lett 174:247–250.

Tress ML, Bodenmiller B, Aebersold R, Valencia A (2008) Proteomics studies confirm the presence of alternative protein isoforms on a large scale. Genome Biol 9:R162. doi:10.1186/gb-2008-9-11-r162

Tyers M, Mann M (2003) From genomics to proteomics. Nature 422:193–197.

Urano K, Kurihara Y, Seki M, Shinozaki K (2010) "Omics" analyses of regulatory networks in plant abiotic stress responses. Curr Opin Plant Biol 13:132–138. doi:10.1016/j.pbi.2009.12.006

Wang R, Fabregat A, Ríos D, et al (2012) PRIDE Inspector: a tool to visualize and validate MS proteomics data. Nat Biotechnol 30:135–137. doi:10.1038/nbt.2112

Wang M, Wang Q, Zhang B (2013) Evaluation and selection of reliable reference genes for gene expression under abiotic stress in cotton (Gossypium hirsutum L.). Gene 530:44–50. doi:10.1016/j.gene.2013.07.084

Wilkinson PA, Winfield MO, Barker GLA, et al (2012) CerealsDB 2.0: an integrated resource for plant breeders and scientists. BMC Bioinformatics 13:219. doi:10.1186/1471-2105-13-219

Wright JC, Sugden D, Francis-McIntyre S, et al (2009) Exploiting proteomic data for genome annotation and gene model validation in Aspergillus niger. BMC Genomics 10:61.

Yachdav G, Kloppmann E, Kajan L, et al (2014) PredictProtein—an open resource for online prediction of protein structural and functional features. Nucleic Acids Res 42:W337–43. doi:10.1093/nar/gku366

Yadav R, Arora P, Kumar S, Chaudhury A (2010) Perspectives for genetic engineering of poplars for enhanced phytoremediation abilities. Ecotoxicology 19:1574–1588. doi:10.1007/s10646-010-0543-7

Yan S, Du X, Wu F, et al (2014) Proteomics insights into the basis of interspecific facilitation for maize (Zea mays) in faba bean (Vicia faba)/maize intercropping. J Proteomics 109:111–124. doi:10.1016/j.jprot.2014.06.027

Yang F, Melo-Braga MN, Larsen MR, et al (2013) Battle through signaling between wheat and the fungal pathogen Septoria tritici revealed by proteomics and phosphoproteomics. Mol Cell Proteomics MCP 12:2497–2508. doi:10.1074/mcp.M113.027532

Yao D, Zhang X, Zhao X, et al (2011) Transcriptome analysis reveals salt-stress-regulated biological processes and key pathways in roots of cotton (Gossypium hirsutum L.). Genomics 98:47–55. doi:10.1016/j.ygeno.2011.04.007

Yu J, Zhao M, Wang X, et al (2013) Bolbase: a comprehensive genomics database for Brassica oleracea. BMC Genomics 14:664. doi:10.1186/1471-2164-14-664

Zhang M, Lv D, Ge P, et al (2014a) Phosphoproteome analysis reveals new drought response and defense mechanisms of seedling leaves in bread wheat (Triticum aestivum L.). J Proteomics 109:290–308. doi:10.1016/j.jprot.2014.07.010

Zhang Y, Cheng Y, Guo J, et al (2014b) Comparative transcriptome analysis to reveal genes involved in wheat hybrid necrosis. Int J Mol Sci 15:23332–23344. doi:10.3390/ijms151223332

Zhao H, Peng Z, Fei B, et al (2014) BambooGDB: a bamboo genome database with functional annotation and an analysis platform. Database (Oxford) 2014:bau006. doi:10.1093/database/bau006

Chapter 3
Omics Approaches to Understand Biotic Stresses: A Case Study on Plant Parasitic Nematodes

Abstract Recently, multiple "omics" sciences such as genomics, transcriptomics, proteomics, and bioinformatics have been introduced. These sciences can be applied and effectively used in the area of biotic stresses, particularly microbial pathogeny studies. In this chapter, we provide deeper insight and discussion on the important applications of the diverse -omics fields of study with a focus on Plant Parasitic Nematodes (PPNs). For every five animal species on the planet, four are nematodes. Almost every animal and plant has at least one parasitic nematode species that is specifically tailored to make use of the food and resources that is represented by the host species, leading to significant losses caused by these pathogens and a need for applicable solutions in order to meet the rapidly growing world population and food demand. While most of the related literature center their focus on transcriptomics, we will be exploring other -omics studies that were published recently, particularly following the completion of the genomes sequences of the three widespread species (*Meloidogyne incognita, M. hapla,* and *Bursaphelenchus xylophilus*) in order to provide an in-depth understanding of major plant biotic stresses, their effects on plants, and the most effective strategies to control them.

Keywords Plant-parasitic nematodes • Biotic stress • Pathogeny • Omics • Bioinformatics • Databases

3.1 Introduction

Pathogens like bacteria, viruses, fungi, parasites, insects and nematodes cause different forms of biotic stresses with their negative effects on plants or any organisms under specific environmental conditions such as geographic location and climate as well as the nature of the particular host. Plants are always exposed to both abiotic (Cramer et al. 2011) and biotic stresses that lead to significant economic losses involving crops threatening agriculture and food security (Check Chap. 1 for more details). The majority of plant and animal diseases are caused by biotic stresses because of the great difficulty in controlling them. Plant reactions or defense approaches to biotic stresses or any other kind of stresses are very complex and involve numerous molecular mechanisms to survive and adapt to the stressful events. In recent years, most studies centered their focus on studying different plant

© The Author(s) 2017

K.A. Mosa et al., *Plant Stress Tolerance*, SpringerBriefs in Systems Biology, DOI 10.1007/978-3-319-59379-1_3

Table 3.1 Different types of biotic stresses on plants; adapted from Allwood et al. (2008)

Interaction	Class definition	Example
Necrotrophic	Pathogens that infect and kill host cells by secreting toxins, degrading enzymes or effector proteins then extract nutrients from the dead host cells	*Septoria tritici* (wheat pathogen)
		Erwinia carotovora (soft rot)
Biotrophic	Pathogens that need to establish feeding relationship with the living host cells in order to live and reproduce	*Magnaporthe grisea* (rice blast disease)
		Pseudomonas syringae (diseases of tomato and bean)
		Basidiomycota (rust fungi)
		Ascomycota (powdery mildew fungi)
Hemibiotrophic	Pathogens that primarily infect living tissues as biotrophic for some time before turning into the necrotrophic phase	*Mycosphaerella graminicola* (wheat pathogen)
		Bipolaris sorokiniana (wheat and barley pathogen)
		Rhynchosporium secalis (scald barley)
		Zymoseptoria tritici (wheat Septoria tritici blotch STB)
Grazer	Pathogens feed on growing plants including herbivorous insects, mammals or other kingdoms	*Phyllotreta nemorum* (flea beetle)
		Pieris brassicae (cabbage white butterfly)
Parasite	Where an organism (parasite) benefits over another organism (host) by deriving its nutrients	*Meloidogyne incognita* (parasitic nematode)

defense strategies against biotic stresses. Plants apply different signaling mechanisms as a response to individual and group stresses (Atkinson and Urwin 2012). Some pathogens are biotrophs, where they infect and feed on a plant but do not kill it. Other pathogens are necrotrophic, where infection and feeding result in the death of the host plant. Another form of pathogens are hemibiotrophs, where the pathogen is initially biotrophic before turning necrotrophic resulting in plant death (Table 3.1).

In the last years and following the revolution of complete genome sequencing projects, bioinformatics and omics disciplines have been introduced that come with many advantages that can be applied to understanding and building deeper knowledge, particularly in regards to complex relationships. Omics technologies can be applied not only for a greater understanding of normal physiological processes, but also for elucidating pathogen/host interactions including life cycles, development, organism resistance and pest survival. Omics technologies are basically targeting the general finding of genes (genomics), mRNA (transcriptomics), proteins (proteomics) and metabolites (metabolomics) in any biological sample (Horgan and Kenny 2011). The inter-relationship between research laboratory techniques with newly developed software and advanced computational tools have guided scientists in answering complex biological questions in terms of bioinformatics. The classical approaches were based on two key concepts; *in vitro* and *in vivo* studies, which were time and cost consuming. On the other hand, with the broad range of omics technologies applications, recent bioinformatics tools and developed computational tools during in silico studies, the answering of difficult research questions has become achievable (Wiwanitkit 2013).

For example, the phenomenon of cross-tolerance allows plants to defend themselves against multiple stresses simultaneously (biotic, abiotic, or in combination). Surprisingly, research on plants exposed to multiple stresses indicate that when a plant becomes resistant to one form of stress, it can develop resistance to other types of stresses as well (Rejeb et al. 2014). One study showed that exposure to mechanical wounding in tomato plants (*Solanum lycopersicum*) can improve their resistance to salt stress by triggering wound-related genes (Capiati et al. 2006). Another study showed that transgenic lines of Arabidopsis with overexpressing genes related to the NLLs (Nictaba-like lectins) from soybean (*Glycine max*) resulted in improvement in the plant tolerance to bacterial infection (*Pseudomonas syringae*), insect infestation (*Myzus persicae*) and salinity (Van Holle et al. 2016). However, more efforts are needed to understand the plant-stress interactions and the signaling pathways among different types of stresses. Hence, the utilization of genomics, transcriptomics and proteomics in this field of microbial pathogeny studies may be beneficial.

3.1.1 Genomics and Transcriptomics of Microbial Pathogens

In basic terms, genomics is the particular type of omics science and bioinformatics that deals with the whole genome with roots in genetics (Check Chap. 2 for more details). There is no doubt that genomic techniques have grown widely to allow numerous applications in the area of microbial pathogeny studies. Databases are tools for collecting genomic information so that it can be easily organized, accessed, managed, updated and retrieved. Normally, searching a database is the first step in omics and bioinformatics approaches. Many databases are accessible and available to the public and some of these can also be applied to pathogeny studies. PubMed, a free resource developed and maintained by the National Center for Biotechnology Information (NCBI), is one of the most widely accessed databases that allows scientists to search for genes and proteins that are related to the pathogenesis of diseases. E-Fungi is a publicly available database that stores integrated data for more than 30 fungal genomes, including genome data, functional annotations and pathway information within a single repository. It also supports comparative analysis driven from the latest comparative genomics studies such as MCL and OrthoMCL cluster analysis. The e-Fungi database can be freely accessed through http://www.e--fungi.org.uk (Hedeler et al. 2007).

Some interesting databases are specialized for specific pathogens. For example, a database focused on bacteria is Bacteriome.org, which is considered the primary resource for the *Escherichia coli* interactome and provides scientists with protein–protein interaction networks and the ability to visualize their structural, functional, and evolutionary relationships (Su et al. 2008). *coli*BASE is another specific database that focuses on *Escherichia coli* and its relatives *Salmonella* and *Shigella* through comparative genomics that can be accessed via http://colibase.bham.ac.uk. *coli*BASE is a relational database (MySQL) comprising of comparative data and provides several analytical tools, such as alignment of the whole genome and lists of putative orthologous genes (Chaudhuri et al. 2004). (See more in Bioinformatics and Web-based analysis in Chap. 3).

Comparative genomics is the process of comparing different genomes using computational techniques. This comparison has become an essential process in omics science. Simply, we can compare the derived genome with the referenced genomes, which are typically allocated in a standard database. The result of the comparison will lead to similarity. Similarity between genomes often means similar genome functions which usually occur in closely related groups of pathogens and newly detected pathogens. For example, comparative genomics was used to clear the confusion over the origin of the swine flu virus in 2009. Ultimately, the virus was discovered to be a new form of virus comprising of a combination of human, swine and avian species (Holmes 2010). Comparative genomics searches can also show how significant the difference between genomes in sequences can be, which is important for classification. The concept of measuring the difference between genomes of species has been used in science for years and is called phylogenetic.

Transcriptomics is the process of identifying and annotating the complete set of transcripts encoded within a genome (Check Chap. 2 for more details). It is the characterization of the gene-expression profiles in different tissues and cell types at a certain developmental stage and/or under defined physiological conditions at a given time. Genome-wide expression profiling at the RNA level has delivered new insight into the transcriptome signatures of biotics stresses or pathogens during infection. The study of the complete set of RNA including messenger RNA, transfer RNA, ribosomal RNA and other non-coding RNAs within a cell or organism provide biologists with important information such as the expressed genes of a genome, regulatory sequences, gene structure, functions and interactions between genes, and identification of candidate genes for any given process or disease when attempting to control such disease. In different tissues or under different physiological conditions or environmental stimuli, the transcriptional response of the genome may change. The discovery of differentially expressed genes is one of the major goals of transcriptome analysis. Over the past 20 years, many techniques have been developed and evolved. Starting from the initial Expressed Sequence Tags (ESTs) along with Serial Analysis of Gene Expression (SAGE), leading to the probe hybridization based method in microarray chips and current RNA-sequencing (RNA-seq) or Next Generation Sequencing (NGS), technologies were developed to monitor the transcriptome rapidly and attain the differentially expressed genes (Mortazavi et al. 2008; Boguski et al. 1994; Schena et al. 1995).

3.1.2 *Proteomics and Metabolomics of Microbial Pathogens*

Proteomics is the specific omics science concerned with employing the techniques of molecular biology, biochemistry, and genetics to study the structures, functions and interactions of the complete set of proteins expressed by a cell, tissue or genome in an organism at a particular time (Check Chap. 2 for more details). In comparison to genomics, the proteomics technique is also widely useful and applicable for studying microbial pathogeny in different ways. Similar to genomics, a protein database is the principal requirement during bioinformatics analysis. PubMed is also an important database for searching proteins. PubMed also provides access to additional relevant databases such as UniProt, PIR, PRF, and PDB. UniProt comprises of three databases that provide the biologist and interested scientists with a complete, excellent and freely accessible resource of protein sequence and functional data (Wu et al. 2006). The first database is Protein Knowledgebase (UniProtKB) that consists of two sections:

1. Swiss-Prot: containing records that are manually annotated with evaluated computational analysis providing a means to search for the function of proteins.
2. Translation of the EMBL nucleotide sequence (TrEMBL): containing records that were automatically annotated and not reviewed with the full manual annotation still pending.

The second database is the UniProt Archive (UniParc), which is a complete and non-redundant database that contains information on the majority of existing protein sequences. The third database is the UniProt Reference Clusters (UniRef) database that combines sets of sequences from the UniProt Knowledgebase (including isoforms) with select UniParc records in clusters. Protein Information Resource-Protein Sequence Database (PIR-PSD) is another free resource of protein informatics that maintains functionally annotated protein sequences. The Proteome database is another information resource that UniProt delivers to scientists and researchers collecting all proteins believed to be expressed only by organisms with completely sequenced genomes. Some of these proteomes have been carefully chosen to be reference proteomes. They include well-studied model organisms and other organisms of interest for microbial pathogeny study and research. For example, *Tobacco mosaic virus* (strain OM) is one of the top ranked for plant viruses that is also included in the proteome database with one protein entry. Another interesting database, 2D–PAGE database (2DBase) is another proteome database of *E. coli* developed to manipulate information obtained by 2D polyacrylamide gel electrophoresis and mass spectrometry (Vijayendran et al. 2007).

Protein is the essential molecule of a cell that participates in certain tasks in presenting the phenotype. Proteins differ from one another largely in their amino acid sequence that is delivered from the nucleic acid sequence of their genes which may result in protein folding to form a specific structure to clarify their activities. Structural proteomics or protein structure prediction is the omics science to predict the structure of a protein which can reveal the pathogenesis process of the pathogen and its interaction with the susceptible and resistant host during the course of a study. Many software tools are available and utilized in protein structure prediction such as online tools including TASSER, SWISS-MODEL, and RaptorX while others are downloadable such as CABS and EasyModeller.

Similar to genomics, comparison between two proteins among different proteomes using computational techniques can indicate the similarity which is commonly termed comparative proteomics. Comparative proteomics is a very useful technique to monitor and analyze proteome variations in response to development, disease or environmental stresses. It is a two-step process where extracts of proteins are first separated to lower the complexity of samples followed by protein identification by mass spectrometry. Difference Gel Electrophoresis (DIGE) is a gel electrophoresis-based technique used for protein fractionation and quantification in protein mixtures that was developed to overcome the limitations of reproducibility and sensitivity in traditional 2D electrophoresis (Minden 2007). The DIGE technique has been continuously improved and remains one of the most important methods in functional proteomics. On the other hand, mass spectrometry is the most commonly used technique to identify and quantify molecules in simple and complex protein mixtures in quantitative proteomics.

Metabolomics is the study of an organism's metabolome to identify and quantify cellular metabolites using analytical technologies, statistical applications and variant methods for information extraction and data interpretation (Check Chap. 2 for more details). Metabolites are the resulting small molecules following metabolism. Metabolites are involved in different functions, including defense and interactions with other organisms. Primary metabolites such as amino acids, ethylene, vitamins, organic acids and nucleotides are synthesized by the cell in order to be involved during growth, development and reproduction. Secondary metabolites such as antibiotics, drugs, fragrances, and pigments are compounds produced by an organism that are not directly involved in primary metabolic processes, though they still have important biological functions (Roessner and Bowne 2009).

Metabolomics is one of the best forms of post-genomic analysis as it can screen changes in metabolites. Metabolic changes are the main characteristic of plant interactions with pathogens, pests and their environment. Metabolomics approaches are commonly used to assess biochemical complexity and challenging scenarios in plant–host interactions by identifying which metabolites originate from the plant host and which are from the interacting pathogen.

Nuclear Magnetic Resonance (NMR) known as NMR spectroscopy is one of the most commonly used techniques in metabolomics. NMR spectroscopy verifies the physical and chemical properties of metabolites. For example, metabolomics based on one- and two-dimensional NMR spectroscopy was used in combination with Principal Component Analysis (PCA), Partial Least Square-Discriminant Analysis (PLS-DA) and Hierarchical Cluster Analysis (HCA) to classify 11 South American Ilex species (Kim et al. 2010). Similarly, a NMR-based metabolomics technique has been successfully used to divulge the metabolic differentiations between five *Verbascum sp.* (mullein) species (Georgiev et al. 2011). Bacterial plant pathogenic interactions are known to cause diseases with economic impact. Dual metabolomics profiling approaches have been used to study metabolite changes between the bacteria *Pseudomonas syringae* and *Arabidopsis thaliana* in order to understand the molecular basis of plant diseases. Along with dual metabolomics, Fourier Transform Infrared (FT-IR) spectroscopy was applied to evaluate the intracellular metabolomes of both host and pathogen under disease and resistance conditions (Allwooda et al. 2010). Raman spectroscopy, Electrospray Ionization (ESI-MS) and Gas Chromatography–Mass Spectrometry (GC-MS) are also commonly used to monitor a biological system to see if differential metabolites exist between control and test material

(Johnson et al. 2003). Ten steps were proposed demonstrating the application of Mass spectrometry-based metabolomics and proteomics approaches to enhance plant resistance to biotic stress (Kushalappa and Gunnaiah 2013).

3.2 A Case Study on Plant Parasitic Nematodes

Nematodes live in diverse environments in both soil and water, spreading from hot spring climates to polar habitats with a very wide range of hosts (including insects to higher mammals). Nematode species can be free-living or parasitic, infecting plants, animals and humans. Nematodes have a simple external morphology that shaped like a worm that is symmetrical and unsegmented. They are essentially aquatic, depending on at least a thin coat of liquid in order to be active and have the ability to spread in a wide range of environments. According to their living and feeding styles, plant parasitic nematodes can be separated into three major types:

1. Ectoparasite: Nematodes usually live outside the plant hosts (and can be easy detected in soil samples) while feeding on roots using their mouth stylets to pierce the root cells in order to draw food and release substances and proteins that aid the nematode in parasitizing their hosts. Examples include *Xiphinema* (dagger nematode), *Longidorus* and *Paralongidorus* (needle nematode) as well as *Criconemella* and *Macroposthhonia* (ring nematode) (Fig. 3.1).
2. Endoparasite: Nematodes usually penetrate the host cell to enter and reside within (and can be easily detected in the tissues samples upon which they feed). There are two different sub-types of endoparasite nematode feeding strategies:

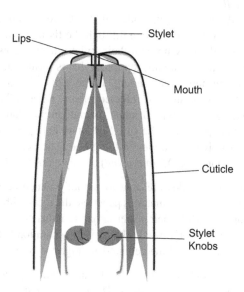

Fig. 3.1 Head morphology of ectoparasites of plant parasitic nematodes, in which the nematode body stays outside the plant host and uses its long mouth stylet to penetrate the cells of the plant roots in order to feed

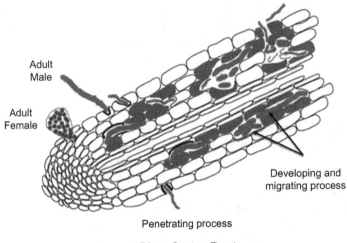

Adult
Male

Adult
Female

Developing and
migrating process

Penetrating process

Plant Cortex Root

Fig. 3.2 Migratory endoparasite of plant parasitic nematodes penetrate, feed, develop, and migrate throughout the root cortex. Juveniles and females migrate to the soil, attack, and enter the healthy host roots. Females lay the eggs in the root cortex. When the eggs hatch, the juveniles either remain in the roots and feed, move to the soil or migrate into the plant

(a) Migratory endoparasite: nematodes spend most of their lifetime migrating via root tissues while destructively feeding and drawing out cytoplasm using their stylet and having no permanent feeding site (Fig. 3.2). Examples of migratory endoparasitic nematodes are *Pratylenchus*, *Radopholus*, and *Hirschmanniella*.

(b) Sedentary endoparasite: nematodes at the J2 stage start penetrating cells near the tip of a root and migrate through the tissue to developing vascular cells. Subsequently, they implant their stylet into the roots and suck up the sap. A gall is the most dramatic symptom that is formed as a result of sedentary endoparasite nematode feeding. Large galls or "knots" can be found throughout the root system of infected plants. Root galls vary in size and shape depending on the type of plant, nematode population levels and species of nematode present in the soil. Sedentary endoparasitic nematodes are the most devastative nematodes in the globe. The two major nematodes in this group are root-knot (*Meloidogyne*) and cyst (*Heterodera*).

3. Semi-endoparasite: nematodes are able to partially penetrate the roots of their hosts. Usually, these types of nematodes penetrate the roots with their head to form permanent feeding sites while their posterior part remains in the soil without movement (e.g. *Rotylenchulus reniformis*). This type of separation becomes more complicated with the occurrence of the nematodes' migratory or sedentary life cycles.

Plant parasitic nematodes have powerful techniques to manipulate their host plants by forwarding and reversing their genetics. Therefore, genome sequencing was the first and most critical step in divulging the essential molecular mechanisms underlying host–parasite interactions in discovering strategies to control plant diseases.

3.3 Importance of Plant-Parasitic Nematodes

The importance of the PPN (Plant Parasitic Nematodes) should not be underestimated as they are a major devastating obligate parasite that infects thousands of plant species. While plant-nematodes feed on all parts of the plant hosts; roots, leaves, seeds, flowers, and stems, the majority of the plant-parasitic nematodes feed on roots to extract the contents (water and nutrients) of plant cells. PPN are considered the invisible enemies of plants as their involvement in growth symptoms is difficult to determine (especially above ground) because these are usually associated with one or more of the abiotic stresses (e.g. temperature stress, water stress, lack of nitrogen etc.), pathogen infections (fungi and bacteria) or in some cases transmitting plant viruses. Infected plants are usually found with spotted patterns as a result of the secretion of pathogen-related proteins (Fig. 3.3) disrupting plant physiology following infection. The majority of plant parasitic nematodes are polyphagous, enabling them to invade many species of plant hosts (e.g. *Aphelenchoides*, *Ditylenchus,* and *Meloidogyne*), while other groups have been developed as host-specific (e.g. *Heteroderidae, Meloidogynidae, Pratylenchidae* and *Neotylenchidae*). Such host-specificity is not only determined by how many species are to be exploited, but also on how the host species are closely related to each other. For example, *Heterodera glycines* (soybean cyst nematode; SCN) cause the greatest crop damage and loss in comparison to other soybean (*Glycines max*) diseases. Table 3.2 lists the economic effects of plant parasitic nematodes in details. When all crops are taken into consideration, the total loss in crop yields caused by plant parasitic nematodes has been estimated to exceed hundreds of billions of US dollars, thus having a significant impact on food security in relation to rapid growths in the human population.

Fig. 3.3 Soybean Cyst Nematode damage, browning and wilted spots are observed in the (soybean) fields that are at risk. One potential issue is that the injury caused by nematodes cannot be diagnosed from observing symptoms unless it reaches the high infestation level. Photo taken by J. Faghih (Faghihi and Ferris, n.d.). Source: https://extension.entm.purdue.edu/fieldcropsipm/insects/nematode.php

Table 3.2 Plant parasitic nematodes of economic importance; adapted from Nicol et al. (2011)

Genus	Primary damage	Plant tissue
Anguina	Seed gall	Seeds, stems, leaves
Bursaphelenchus	Wilt	Seeds, stems, leaves
Criconemella	Ring	Roots
Ditylenchus	Stem and bulb	Stems, leaves
Globodera	Cyst	Roots
Helicotylenchus	Spiral	Roots
Heterodera	Cyst	Roots
Hirschmanniella	Root	Roots, tubers
Hoplolaimus	Lance	Roots
Meloidogyne	Root-knot	Roots
Pratylenchus	Lesion	Roots
Radopholus	Burrowing	Roots, tubers
Rotylenchulus	Reniform	Roots
Scutellonema	Spiral	Roots, tubers
Tylenchulus	Citrus	Roots

3.4 Omics Studies Related to Plant Parasitic Nematodes

Plant parasitic nematodes were observed in early history to be major agricultural pathogens and to cause serious plant diseases. As the full extent of the damage was recognized by biological and agricultural scientists, many studies conducted on plant parasitic nematodes became increasingly important to understand their biology. The purpose was to build a deeper detailed knowledge of the causative agents of plant diseases, including the nematode/host interaction, life cycle, development, plant resistance and pest survival.

C. elegans (*Caenorhabditis elegans*) was the first nematode to have its complete genome sequenced in 2002. Since then, it has become the model of choice and used in many studies conducted towards improving our understanding of the basic biology of plant parasitic nematodes. However, having many different characteristics have made them highly difficult to understand. Issues have included the wide range of diversity in some species, different developmental stages (particularly inside the roots) and the difficulty of having them cultured.

3.4.1 Genomics of Nematode

Access to whole genome sequencing provides a methodology that affords a comprehensive view in analyzing all genes while avoiding the difficulty of finding silenced genes or those expressed at low levels by using other strategies. However, whole genome sequencing may still provide an incomplete picture as a result of having difficulty in predicting certain transcripts and particular regions that are to be sequenced in addition to the complex biological diversity found in nematodes.

Existing tools and techniques have been improved while others have emerged in recent years. These include Next Generation Sequencing (NGS) (Roche, Branford, CT, USA), Illumina sequencing (Illumina Inc., San Diego, CA, USA) and Sequencing Oligonucleotide Ligation and Detection (SOLiD) (Applied Biosystems, Carlsbad, CA, USA), which are all new sequencing technologies that can generate sequence data at massively accelerated speeds and at very low cost compared to the typical method of Sanger sequencing. Despite the tremendous reduction in costs and the increased data access and availability, considerable challenges remain. In particular, issues may be related to data analysis and interpretation and other specific problems such as assembly process and data storage when using NGS technology.

All related parasitic roundworm (nematodes) studies have confirmed that nematodes are the major cause of human, animal and plant diseases resulting in major socio-economic impacts globally. Therefore, newly available information from functional genomics, transcriptomics and proteomics can greatly help in developing more efficient parasitic nematode control programs. *Caenorhabditis elegans* (*C. elegans*) was the first multicellular-organism (animal) to have a completed genome sequence (*C. elegans* Sequencing Consortium, 1998), which allowed scientists to divulge answers for questions previously unimaginable. The sizes of nematodes genomes that have been analyzed and assembled to date range between 19.67 Mb in *Pratylenchus coffeaecarries* (as the smallest nematode genome) and as high as 443.017 Mb in *Oesophagostomum dentatum*. Table 3.3 lists the number of nematode species to date that have been completely sequenced and published. This includes four species of plant parasitic nematodes; two from the genus Meloidogyne (sedentary endoparasites), *Meloidogyne incognita* and *Meloidogyne hapla*, with the other two being migratory endoparasites, *Bursaphelenchus xylophilus* (pine wood nematode) with a genome size of 73.09 Mb and *Globodera pallida* with a genome size of 123.63 Mb. Efforts engaged in the sequencing of plant parasitic nematode genomes and the related data that has been recently published have revealed important distinctions that can suggest biological differences between various nematode species, including species from similar genus such as *M. incognita* and *M. hapla* (Table 3.4). The major genomic differences between both these species are the small size of the *M. hapla* genome at 54 Mb in comparison to *M. incognita* which totals 86 Mb. While *M. incognita* is an asexually reproducing nematode (polyploidy and/ or aneuploidy), *M. hapla* has a meiotic reproduction lifestyle. Further genome sequencing projects in different nematode species are in progress, including *Meloidogyne arenaria*, *Globodera rostochiensis* and *Heterodera glycines* from plant parasitic nematodes.

Bursaphelenchus xylophilus or pine wood nematode is the major pathogen responsible for pine wilt disease in Asia and Europe that causes severe damage to several species of pine trees. The genome sequencing of *B. xylophilus* provides a unique opportunity to understand disease mechanisms as well as plant parasitism in efforts to control the devastating disease it causes. *B. xylophilus* has a composite life cycle that combines fungal feeding and plant parasitism with insect-associated stages. The genome sequence of this species comprises of 18,074 genes spread across six chromosomes, which would provide an excellent resource in understanding the biology of this unusual parasite.

Table 3.3 List of nematodes with completed genomes; adapted from Martin et al. (2014)

Published genomes	NCBI taxon id	Publication link	Publication year
Ascaris suum	6253	Ascaris suum draft genome	2011
Brugia malayi	6279	Draft Genome of the Filarial Nematode Parasite Brugia malayi	2007
Bursaphelenchus xylophilus	6326	http://journals.plos.org/plospathogens/article?id=10.1371/journal.ppat.1002219	2011
Caenorhabditis briggsae	6238	The Genome Sequence of Caenorhabditis briggsae: A Platform for Comparative Genomics	2003
Caenorhabditis elegans	6239	Genome Sequence of the Nematode *C. elegans*: A Platform for Investigating Biology	1998–2002
Dirofilaria immitis	6287	The genome of the heartworm, Dirofilaria immitis, reveals drug and vaccine targets	2012
Globodera pallida	36090	The genome and life-stage specific transcriptomes of Globodera pallida elucidate key aspects of plant parasitism by a cyst nematode	2014
Haemonchus contortus	6289	The genome and transcriptome of Haemonchus contortus, a key model parasite for drug and vaccine discovery	2013
Heterorhabditis bacteriophora	37862	A Lover and a Fighter: The Genome Sequence of an Entomopathogenic Nematode Heterorhabditis bacteriophora	2013
Loa loa	7209	Genomics of Loa loa, a Wolbachia-free filarial parasite of humans	2013
Meloidogyne hapla	6305	Sequence and genetic map of Meloidogyne hapla: A compact nematode genome for plant parasitism	2008
Meloidogyne incognita	6306	Genome sequence of the metazoan plant-parasitic nematode Meloidogyne incognita	2008
Necator americanus	51031	Genome of the human hookworm Necator americanus	2014
Pristionchus pacificus	54126	The Pristionchus pacificus genome provides a unique perspective on nematode lifestyle and parasitism	2008
Romanomermis culicivorax	13658	The genome of Romanomermis culicivorax: revealing fundamental changes in the core developmental genetic toolkit in Nematoda	2013
Trichinella spiralis	6334	The draft genome of the parasitic nematode Trichinella spiralis	2011
Trichuris muris	70415	Whipworm genome and dual-species transcriptome analyses provide molecular insights into an intimate host-parasite interaction	2014
Trichuris suis	68888	Genome and transcriptome of the porcine whipworm Trichuris suis	2014
Trichuris trichiura	36087	Whipworm genome and dual-species transcriptome analyses provide molecular insights into an intimate host-parasite interaction	2014

Table 3.4 General genomic features for different nematode species; adapted from Abad et al. (2008)

	B. xylophilus	M. incognita	M. hapla	C. elegans	P. pacificus	B. malayi
Estimated size of genome (Mb)	63–75[a]	47–51	54	100[a]	172.5	88.3
Chromosome number	6	Variable	16	6	6	6
Total size of assembled sequence (Mb)	74.6	86	53	100	172.5	95.8
Number of genes	18,074	19,212	14,420	20,416	24,216	11,590
Number of proteins	18,074	20,365	13,072	24,890	24,239	14,298
Average protein length	345	354	309.7	439.6	331.5	311.9
Gene density (genes per Mb)	242.3	223.4	270	249	140.4	221.8

[a]The genome size was estimated using real-time PCR

At first, little was known about the molecular basis of the interactions between nematodes and their host plants. In the last few years and after complete genome projects became more applicable in terms of time and costs, a series of advances were made and enormous amounts of data have become available. This includes information regarding chromosome number and structure, genome size, genes behaviors, protein coding gene prediction, functional annotation and more.

3.4.2 Nematode Transcriptomics Analysis

Many plant parasitic nematode transcriptome databases have been constructed and annotated as a rapid and cost-effective approach for gene identification. However, despite the obvious benefits from recent advances in next-generation sequencing technologies as the most effective technique for measuring the transcriptomes of organisms (Berglund et al. 2011), the older technique of utilizing DNA microarrays is still in use. Transcriptomic approaches have included the generation of Expressed Sequence Tags (ESTs) from different nematode lifecycle stages or infected plant tissues and have become the most important strategy for studying plant parasitic nematodes at the molecular level.

While complete genome sequencing provides us with insight into what the organism is capable of, ESTs provide us with the knowledge of which parts are functioning in which tissue during which period of time and under what conditions. ESTs are considered as validated approaches in discovering novel genes which makes them a crucial tool for parasitic nematode researchers. ESTs have been used mainly for expression profiling, interpreting proteomics, phylogenetic analysis and DNA amplification. The majority of EST studies have been focused on understanding the mechanisms (mainly protein effectors) involved in plant-parasitism caused by plant parasitic nematodes. Some of these studies have been conducted to compare different environmental studies at different nematode species' stages in different nematode strains or species. For example, a microarray study to determine the genetic expression signature of two different *H. glycines* populations (compatible population and incompatible population) after being exposed to the resistant *Glycine max* genotype (Peking) was conducted at two different stages (pre-parasitic and post-infection). The results identified numerous putative parasitism genes that are differentially expressed between the two populations. Therefore, the results indicate how *H. glycines* may develop different mechanisms to overcome resistance. Next-generation sequencing (NGS) technology has made it possible to generate large amounts of genome and transcriptome sequencing data and relevant applications in a few days at a fraction of the cost of traditional Sanger-sequencing, allowing unparalleled exploration of high nematode diversity. However, assembling and annotating transcriptomic data into genomic resources remains a challenge because of the short reads, the quality issues in these kinds of data as well as the presence of contaminants in uncultured samples.

With the growing demand for drug discovery and vaccine development, the potential exists to develop resistance to common parasites especially those causing large economic losses around the world. *Haemonchus contortus* is one of the major infective nematodes of ruminants and one of the most widely used in resistance research (Schwarz et al. 2013). The assembled and annotated draft genome along with the comprehensive transcriptomic sequencing and analysis of *H. contortus* was published in 2013. The draft sequence was assembled using a collection of different sequencing technologies including 454 Sequencing Systems using shotgun reads in combination with paired end reads as well as paired-end sequencing from an Illumina HiSeq in order to generate accurate draft assemblies. TopHat software was used to map transcriptome reads against *C. elegans* (reference genome). A *de novo* gene model was used to curate and predict 21,799 protein-coding genes with very low (7%) gene density against 20,532 of *C. elegans* with 30% gene density. Six life stages of *H. contortus* were sampled to analyze their gene expressions using RNA-seq technology. 11,295 genes were found to be significantly up or down regulated through all six life stages. Further analysis has also been conducted in this area such as functional annotation, gene ontology, divulging metabolic pathways and understanding drug metabolism etc. (Schwarz et al. 2013).

3.4.3 Proteomics of Plant Parasitic Nematodes

The use of proteomics in the study of plant parasitic nematodes had initially been a struggle due to shortcomings of the sequenced genome or ESTs data. However, following the revolution of genomics and transcriptomics techniques, proteomics became possible and the next step in studying plant nematode biological systems and their interactions. Many new techniques have been improved to advance proteomics studies in nematodes, including Mass Spectrometry (MS), Multidimensional Liquid Chromatography (MudPIT), Two-Dimensional gel Electrophoresis (2-DE) as well as recent advancements in bioinformatics tools. Plant parasitic nematodes can be either ectoparasites or endoparasites, which can be sedentary or migratory. In the sedentary case, the female initiates the permanent nematode feeding site called giant-cells by establishing a tunnel (stylet) into the host root and becoming immobile afterwards. Nematodes that utilize this kind of feeding strategy develop complex interactive relationships with host cells. Understanding the signal exchange that occurs during infection of plants is complicated and the alteration of many gene expressions involved in feeding site formation remains a mystery. To date, with the powerful new techniques many studies have been conducted resulting in significant progress in identifying the nematode-responsive plant genes.

As a result of having the complete genome of the root-knot nematode *Meloidogyne hapla* sequenced and annotated, a database named HapPep3 was created with the complete proteome of this plant-parasite including 14,420 proteins. The database was made available at the repository of model organism proteomes (http://supfam. mrc-lmb.cam.ac.uk/SUPERFAMILY). The use of LC/MS analysis including comparisons of gene ontology, protein domains, signaling and localization predictions to *M. hapla* protein extracts resulted in a total of 516 non-redundant proteins being identified. The referenced pathways in the free-living nematode (*Caenorhabditis elegans* as a model) were examined to determine their similarity to *M. hapla*. Using 2-DE gels, the complete proteomic map has been compiled for *Heterodera glycines* (a soybean cyst nematode) and the identified proteins were characterized by their functions. *Heterodera glycines* demonstrated 816 different proteins as well as 20 secreted proteins including pectatelyase, β-1,4- endoglucanase, and expansin, which are involved in plant cell wall degradation. The authors also used 2D–PAGE to analyze gene expression at the translational level and evaluated the identified protein function using the Gene Ontology (GO) database (Chen et al. 2011).

3.4.4 Bioinformatics and Web-Based Analysis Tools

Many international organizations and groups such as the International Federation of Nematology Societies or IFNS (http://www.ifns.org/) and the Society of nematologists or SON (https://nematologists.org/) were formed and developed to exchange information, hold regular meetings and bring nematologists together to serve the purpose of promoting and extending knowledge in the science of nematology in all respects. This section aims to provide nematologists in worldwide (national and regional) nematology societies with many of the online websites dedicated to nematode sequence analysis in order to improve the overall awareness of nematodes and the science of nematology. Currently, some of these sites are specializing and working as web-based interfaces for developed databases that provide fast and easy access to the available information on the parasitic nematodes and store the results of various conducted analyses.

Nematode.net was built in 2000 as part of the project "A Genomic Approach to Parasites from the Phylum Nematoda" funded by the National Institute of Allergy and Infectious Diseases (NIAID) as an effort of the Genome Sequencing Center at Washington University (GSC) joined together with a consortium including the Nematode Genomics group in Edinburgh and the Pathogen Sequencing Unit in the Sanger Institute to provide an accessible and searchable repository that allows the scientist community to integrate new sequence data from parasitic nematodes. Nematode.net (accessed via http://www.nematode.net) is one of a collection of two databases that falls under the Helminth.net umbrella offering a wide range of services and resources. Nemaode.net provides rapid access to omics data for almost 84 species of parasitic roundworms (nematodes) causing diseases in humans, animals and plants with socio-economic importance. This site allows browsing, searching and retrieving of EST clusters and contig information generated from cDNA clones as well as some by PCR using the nematode transplice leader SL1. This website provides a number of useful tools in order to access and analyze its database. NemaBLAST is one of the tools integrated to perform BLAST searches by clade, species and library to screen for the existence or absence of a query across a spectrum of organisms.

The AmiGo tool allows users to access and search the Gene Ontology (GO) database associated with EST clusters to browse and query for gene products that are involved in similar biological processes, molecular function and cellular components—more documentation and information are available at http://www.geneontology.org/. NemaPath is one of the nematode.net components that provides information about the enzymatic pathways of curated nucleotide sequences (cDNAs, ESTs and reference sequences) in given nematodes by comparing it to the annotated protein sequences in the KEGG (Kyoto Encyclopedia of Genes and Genomes) database (Fig. 3.4). It also offers direct comparison between any two organisms in the database.

The nematode protein family (NemFam) tool provides access to a database of conserved regions of nematode-related proteins that are not stored in other protein databases. The Nemfam database was built with over 214,000 polypeptides from 32 different nematode species, 27 of which are parasites of vertebrates or plants. The identification number assigned to each protein family can be used to access a gbrowse to identify the features of each family member and to describe what annotations are associated with it. NemaGene is a data mining tool that allow users to access the NemaGene database and search for transcripts or to explore transcript expression by name, contig id or life cycle stage in a given organism. The NemaGene database is a collection of all transcript assembly contigs (using both the Sanger clustering method and 454 cDNA) processed at the Genome Institute. NemaGene EST clusters were built from identical and semi-identical contigs and then brought together to form clusters (where cluster members were obtained from the same gene or from multigene families with highly identical sequences). Clusters were annotated using InterPro (a protein families database) where conserved domains and functional sites were identified. Clusters were then associated with the Gene

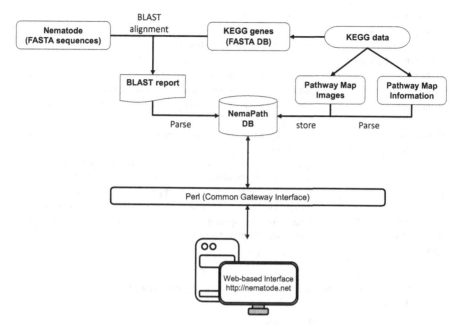

Fig. 3.4 The pipeline workflow of NemaPath. Alignment of EST associations to the KEGG database. Adapted from Wylie et al. (2008)

Ontology (GO) and Kyoto Encyclopaedia of Genes and Genomes (KEGG) databases. Currently, NemaGene EST clusters are combined with a *C. elegans* (a nematode organism model) database from WormBase in order to find any relationships between the queries and across multiple datasets.

NemaBase is a database that was built as a component of Nemaplex (Nematode-Plant Expert Information System) to offer a wide range of services and resources. NemaBase allows users to access available information on plant and nematode interactions (Table 3.5). Currently, the NemaBase database hosts more than 55,000 entries with data on more than 6400 plant species and 1109 nematode species (Parkinson et al. 2004). WormBase (accessed via http://www.wormbase.org/) is a web-accessible database that is dedicated to provide the research community with the most accurate and up-to-date searchable biology, genetics and genomics information concerning the nematode model organism *Caenorhabditis elegans* and other related nematodes. The database is updated regularly with a new release every 2 months. It comprises of different sets of data such as the reference genome sequence of *C. elegans* and other *Caenorhabditis* genomes, curated annotations, gene families, genetic maps, markers and other information. The WormBase web-based interface provides users with many tools to access, search and retrieve data from its database. As of June 1st 2016, WormMart (the WormBase primary data mining facility) has been replaced by WormMin, which maintains existing functionalities with more powerful and useful features (Table 3.6).

Table 3.5 Search options for Nemabase (retrieved from: http://plpnemweb.ucdavis.edu/nemaplex/Nemabase2010/Nemabase%20Search%20Menu.htm)

Search options	Query description
Nematode host range	Search for the host range of a nematode species, including the level of susceptibility or resistance
Host status of plants to nematodes	Search for cultivars or varieties of plants that meet specified characteristics in their host status to nematodes
Plant host status	Search for the host status of a plant to all nematode species
Plant family	Search for the host status of plants in a specified family to feeding nematodes
Plants resistant to a nematode genus/species	Search for all plants that are resistant or non-hosts to a Nematode genus/species
Plant usage	Search for plants with a specified usage that are non-hosts or have some level of resistance to a specified nematode species
Plant type	Search for plants with a specified growth habit that are non-hosts or have some level of resistance to a specified nematode species
Plants that are non-hosts or resistant to a nematode species	Search for plant species, cultivars or rotation of cover crops with resistance to one nematode species

Table 3.6 Variety of bioinformatics and data analysis tools provided by WormBase (source: http://www.wormbase.org/tools#0--10)

Tool	Description
[a]Blast/Blat	**BLAST** finds regions of similarity between biological sequences by comparing nucleotide or protein sequences to sequences in the WormBase database according to precomputed parameters
	BLAT finds DNA sequences of 95% and greater similarity of a length of 25 bases or more, and finds protein sequences of 80% and greater similarity of a length of 20 amino acids or more
[a]GBrowse	It allows users to browse through the whole genome for tracks corresponding to specific genome regions. Furthermore, it allows a comparative view of two genomes side by side
Genetic map	It depicts which chromosome contains the gene and precisely where the gene is located on that chromosome as well as the distances between genes
Protein/nucleotide aligner	This tool displays the alignment of various types of data arranging the sequences of DNA, RNA or protein to identify regions of similarity in a genomic context
[a]Ontologies	It is a public ontology browser including:
	– Gene ontology (biological process, cellular component and molecular function)
	– Anatomy ontology
	– Human disease ontology
	– Worm life stage ontology
	– Phenotype ontology
Queries	Query language searches underlying the ACeDB database using:
	WormBase Query Language (WQL) or Ace Query Language (AQL)
Textpresso	Allows to search for papers with full text on *C. elegans*
Tissue enrichment analysis	It finds tissues that are representative regarding gene expression annotation frequency and *C. elegans* tissue ontology
[a]WormMine	Data mining platform for *C. elegans* and related nematodes

[a]Most used tools

References

Abad P., Gouzy J., Aury J.-M., Castagnone-Sereno P., Danchin E. G. J., Deleury E., ... Wincker, P. (2008). Genome sequence of the metazoan plant-parasitic nematode *Meloidogyne incognita*. *Nature Biotechnology*, 26, 909–915

Allwood, J. W., Ellis, D. I., & Goodacre, R. (2008). Metabolomic technologies and their application to the study of plants and plant-host interactions. *Physiologia Plantarum*, *132*(2), 117–135. doi:10.1111/j.1399-3054.2007.01001.x

Allwooda, W., Clarke, A., Goodacreb, R., & Mura, L. A. J. (2010). Dual metabolomics: A novel approach to understanding plant–pathogen interactions. *Phytochemistry*, *71*(5), 590–597. doi:10.1016/j.phytochem.2010.01.006

Atkinson, N. J., & Urwin, P. E. (2012). The interaction of plant biotic and abiotic stresses: from genes to the field. *Journal of Experimental Botany*, *63*(10), 3523–3543. doi:10.1093/jxb/ers100

Berglund, E. C., Kiialainen, A., Syvänen, A.-C., Sanger, F., Nicklen, S., Coulson, A. R., ... Budowle, B. (2011). Next-generation sequencing technologies and applications for human genetic history and forensics. *Investigative Genetics*, *2*(1), 23. doi:10.1186/2041-2223-2-23

Boguski, M. S., Tolstoshev, C. M., & Bassett, D. E. (1994). Gene discovery in dbEST. *Science (New York, N.Y.)*, *265*(5181), 1993–4. doi:10.1126/science.8091218

Capiati, D. A., País, S. M., & Téllez-Iñón, M. T. (2006). Wounding increases salt tolerance in tomato plants: Evidence on the participation of calmodulin-like activities in cross-tolerance signalling. *Journal of Experimental Botany*, *57*(10), 2391–2400. doi:10.1093/jxb/erj212

Chaudhuri, R. R., Khan, A. M., & Pallen, M. J. (2004). coliBASE: an online database for Escherichia coli, Shigella and Salmonella comparative genomics. *Nucleic Acids Research*, *32*(Database issue), D296–D299. doi:10.1093/nar/gkh031

Chen, X., M. H. MacDonald, F. Khan, W. M. Garrett, B. F. Matthews, and S. S. Natarajan. (2011). Two-dimensional proteome reference maps for the soybean cyst nematode Heterodera glycines. *Proteomics*, 11:4742–4746.

Cramer, G. R., Urano, K., Delrot, S., Pezzotti, M., & Shinozaki, K. (2011). Effects of abiotic stress on plants: a systems biology perspective. *BMC Plant Biology*, *11*(1), 163. doi:10.1186/1471-2229-11-163

Faghihi, J., & Ferris, V. (n.d.). *Soybean cyst nematode, pests, soybean.* Integrated Pest Management, IPM Field Crops, Purdue University. Retrieved May 31, 2017, from https://extension.entm.purdue.edu/fieldcropsipm/insects/nematode.php

Georgiev, M. I., Ali, K., Alipieva, K., Verpoorte, R., & Choi, Y. H. (2011). Metabolic differentiations and classification of Verbascum species by NMR-based metabolomics. *Phytochemistry*, *72*(16), 2045–2051. doi:10.1016/j.phytochem.2011.07.005

Hedeler, C., Wong, H. M., Cornell, M. J., Alam, I., Soanes, D. M., Rattray, M., ... Paton, N. W. (2007). e-Fungi: a data resource for comparative analysis of fungal genomes. *BMC Genomics*, *8*, 426. doi:10.1186/1471-2164-8-426

Holmes, E. C. (2010). The comparative genomics of viral emergence. *Proceedings of the National Academy of Sciences of the United States of America*, *107* (Suppl8), 1742–6. doi:10.1073/pnas.0906193106

Horgan, R. P., & Kenny, L. C. (2011). SAC review " Omic " technologies: proteomics and metabolomics. *The Obstetrician & Gynaecologist*, *13*, 189–195. doi:10.1576/toag.13.3.189.27672

Johnson, H. E., Broadhurst, D., Goodacre, R., & Smith, A. R. (2003). Metabolic fingerprinting of salt-stressed tomatoes. *Phytochemistry*, *62*(6), 919–928. doi:10.1016/S0031-9422(02)00722-7

Kim, H. K., Khan, S., Wilson, E., Kricun, S. P., Meissner, A., Goraler, S., ... Verpoorte, R. (2010). Metabolic classification of South American Ilex species by NMR-based metabolomics. *Phytochemistry*, *71*(7), 773–784. doi:10.1016/j.phytochem.2010.02.001

Kushalappa, A. C., & Gunnaiah, R. (2013). Metabolo-proteomics to discover plant biotic stress resistance genes. Trends in Plant Science, 18(9), 522–531.

Martin, J., Rosa, B. A., Ozersky, P., Hallsworth-Pepin, K., Zhang, X., Bhonagiri-Palsikar, V., ... Mitreva, M. (2014). Helminth.net: expansions to Nematode.net and an introduction to Trematode.net. *Nucleic Acids Research, 43*(D1), D698–D706. doi:10.1093/nar/gku1128

Minden, J. (2007). Comparative proteomics and difference gel electrophoresis. *BioTechniques, 43*(6), 739–745. doi:10.2144/000112653

Mortazavi, A., Williams, B. A., McCue, K., Schaeffer, L., & Wold, B. (2008). Mapping and quantifying mammalian transcriptomes by RNA-Seq. *Nature Methods, 5*(7), 621–628. doi:10.1038/nmeth.1226

Nicol, J. M., Turner, S. J., Coyne, D. L., Nijs, L. Den, & Hockland, S. (2011). Current Nematode Threats to World Agriculture. In Genomics and Molecular Genetics of Plant-Nematode Interactions (pp. 21–43). Springer. http://doi.org/10.1007/978-94-007-0434-3

Parkinson, J., Whitton, C., Schmid, R., Thomson, M., & Blaxter, M. (2004). NEMBASE: a resource for parasitic nematode ESTs. *Nucleic Acids Research, 32*(Database issue), D427–D430. doi:10.1093/nar/gkh018

Rejeb, I., Pastor, V., & Mauch-Mani, B. (2014). Plant Responses to Simultaneous Biotic and Abiotic Stress: Molecular Mechanisms. *Plants, 3*(4), 458–475. doi:10.3390/plants3040458

Roessner, U., & Bowne, J. (2009). What is metabolomics all about? *BioTechniques, 46*(5 SPEC. ISSUE), 363–365. doi:10.2144/000113133

Schena, M., Shalon, D., Davis, R. and Brown, P. (1995). Quantitative Monitoring of Gene Expression Patterns with a Complementary DNA Microarray. *Advancement of Science, 270*(5235), 467–470.

Schwarz, E. M., Korhonen, P. K., Campbell, B. E., Young, N. D., Jex, A. R., Jabbar, A., ... Gasser, R. B. (2013). The genome and developmental transcriptome of the strongylid nematode Haemonchus contortus. *Genome Biology, 14*(8), R89. doi:10.1186/gb-2013-14-8-r89

Su, C., Peregrin-Alvarez, J. M., Butland, G., Phanse, S., Fong, V., Emili, A., & Parkinson, J. (2008). Bacteriome.org—An integrated protein interaction database for E. coli. *Nucleic Acids Research, 36*(Suppl. 1), 632–636. doi:10.1093/nar/gkm807

Van Holle, S., Smagghe, G., & Van Damme, E. J. M. (2016). Overexpression of *Nictaba*-like lectin genes from Glycine max confers tolerance toward *Pseudomonas syringae* infection, aphid Infestation and salt stress in transgenic *Arabidopsis* plants. *Frontiers in Plant Science, 7*(October), 1590. doi:10.3389/fpls.2016.01590

Vijayendran C, Burgemeister S, Friehs K, Niehaus K, Flaschel E. (2007). 2DBase: 2D-PAGE database of Escherichia coli. *Biochemical and Biophysical Research Communications, 363*(3), 822–827. doi:10.1016/j.bbrc.2007.09.050

Wiwanitkit, V. (2013). Utilization of multiple "omics" studies in microbial pathogeny for microbiology insights. *Asian Pacific Journal of Tropical Biomedicine, 3*(4), 330–333. doi:10.1016/S2221-1691(13)60073-8

Wu, C. H., Apweiler, R., Bairoch, A., Natale, D. A., Barker, W. C., Boeckmann, B., ... Suzek, B. (2006). The Universal Protein Resource (UniProt): an expanding universe of protein information. *Nucleic Acids Research, 34*(Database issue), D187–91. doi:10.1093/nar/gkj161

Wylie, T., Martin, J., Abubucker, S., Yin, Y., Messina, D., Wang, Z., ... & Mitreva, M. (2008). NemaPath: online exploration of KEGG-based metabolic pathways for nematodes. *BMC Genomics, 9*(1), 525.

Chapter 4
Functional Genomics Combined with Other Omics Approaches for Better Understanding Abiotic Stress Tolerance in Plants

Abstract Recent development in omics technologies such as genomics, proteomics, transcriptomics, metabolomics as well as functional genomics has helped the identification and functional characterization of a large number of gene families and mechanisms associated with abiotic stress tolerance in plants. In this chapter, we will mainly discuss gene expression and the regulatory network in response to different abiotic stresses, as well as the forward and reverse genetic approaches used to elucidate the function of abiotic stress tolerance genes. Additionally, special attention will be given to metabolomics and ionomics as powerful tools for understanding abiotic stress tolerance in plants.

Keywords Functional genomics • Abiotic stress • Metabolomics • Ionomics • Forward genetics • Reverse genetics • Overexpression • RNAi • VIGS • T-DNA • Mutagenesis

4.1 Introduction

Abiotic stresses caused by different environmental factors such as salinity, drought, high temperature, nutrient deficiency and toxicity greatly challenge crop sustainability and limit the quantity and quality of the cultivated plants. According to the UN DESA report, "World Population Prospects: The 2015 Revision", the world population is expected to reach 8.5 billion and 9.7 billion by the years 2030 and 2050 respectively (http://www.un.org/en/development/desa/news/population/2015-report.html). Therefore, current agricultural production rates may need to be doubled by 2050 in order to fulfill the augmented demand for food as a result of the increased population (Ray et al. 2013). Worldwide plant breeding programs have been developed to enhance plant tolerance to various abiotic stresses. Progress in molecular biology and omics technologies, in addition to functional genomics, has revealed a huge number of gene families and mechanisms associated with improved plant productivity and tolerance to abiotic stresses.

In this chapter, we will provide an overview of the functional genomics advances merged with other omics to investigate the abiotic stress tolerance mechanisms in plants. We will emphasize on the regulatory mechanisms of some genes and transcription factors involved in different abiotic stresses. We will then discuss the

© The Author(s) 2017
K.A. Mosa et al., *Plant Stress Tolerance*, SpringerBriefs in Systems Biology,
DOI 10.1007/978-3-319-59379-1_4

forward and reverse genetic strategies used to illustrate the abiotic stress tolerance mechanism in plants, supported by some examples of transgenic approaches utilized to cope with these stresses. We will also highlight the potential of ionomics analysis to help reveal the mechanisms of elemental or ion transport, uptake, exclusion, and compartmentalization. Additionally, the metabolomics approaches for the identification of potential metabolites associated with abiotic stress tolerance will be discussed.

4.2 Conceptual Framework for Employing Omics Technologies to Understand the Abiotic Stress Tolerance

Recent advances in omics technologies and its application to plant sciences has opened new horizons by merging different omics approaches like genomics, transcriptomics and metabolomics with functional genomics approaches to elucidate mechanisms involved in abiotic stress tolerance. A proposed conceptual framework is illustrated in Fig. 4.1 to demonstrate the integration of omics with functional

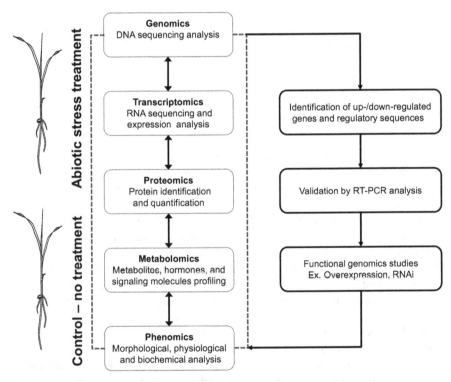

Fig. 4.1 Genomics, transcriptomics, proteomics, metabolomics and phenomics are used to identify upregulated and/or downregulated genes and regulatory sequences from the target plant species under the tested abiotic stress. These genes and regulatory sequences could be the target for functional genomics studies using gain of function (such as overexpression) or loss of function (such as RNAi) approaches to elucidate the roles of candidate genes

genomics strategies in plant abiotic stress tolerance analysis. Genomics analysis can be employed to obtain large-scale sequences of the target plant species genome (Türktaş et al. 2015). Additionally, at the laboratory scale, plants can be treated with a particular abiotic factor (drought, salt, cold etc.) and then compared with non-treated controls. Both the treated and non-treated plants can be subjected to transcriptomics analysis to identify the transcripts of upregulated and downregulated responsive genes and regulatory sequences, which can then be validated by RT-PCR analysis (Fig. 4.1). The integration of metabolomics with transcriptomics and proteomics can help link metabolites with their producer genes, as well as allow the mapping of these metabolites/genes into metabolic pathways (Kumar A et al. 2014). Details about plant metabolic pathways can be harnessed from the plant-specific metabolic pathway database (PlantCyc: http://pmn.plantcyc.org/PLANT/organism-summary?object=PLANT). Metabolomics and proteomics analysis should be performed for the tested samples of the target cells, tissues, and organelles, under the tested abiotic factor (drought, salt, cold, ... etc), as the quality and quantity of the metabolome and proteome are varied spatially and temporally (Kushalappa and Gunnaiah 2013). In contrast, phenomics, which is the high throughput analysis of plant physiology (Furbank and Tester 2011), can be integrated with all other omics to identify candidate genes involved in abiotic stresses. Identified genes by all the above-mentioned omics technologies can be then functionally characterized using functional genomics approaches like gene overexpression or silencing and transgenic plants can be compared with wild-type plants (Fig. 4.1). Phenomics can be used in this case to understand the mechanisms involved in the observed phenotype and correlate them with the tested form of stress.

4.3 Gene Expression and Regulatory Network in Response to Abiotic Stress

Unlike animals, plants are unable to move away from unfavorable environmental abiotic stresses such as salt, drought, high temperature, cold etc. Hence, they develop adaptation mechanisms at the physiological, biochemical and molecular levels regulated by stress-responsive gene expression to survive and grow under these harsh environmental conditions. Transcription factors (TFs) play major roles in regulating gene expression in response to environmental abiotic stresses. Different families of plant TFs have been identified based on their DNA-binding domains (DBDs) such as Abscisic Acid (ABA) Response Elements (ABREs), Drought Reacting Element Binding proteins (DREBs), Basic Region/leucine Zipper (bZIP), Ethylene Response Factors (ERFs) and WRKY (Yamasaki et al. 2013; Nakashima et al. 2014).

Several stress related genes might be induced and regulated by ABA, a phytohormone which is produced when plants are exposed to stresses; this hormone helps in tightening the stomata to prevent water loss under water deficiency conditions. However, not all stress-induced genes are triggered by ABA (Shinozaki et al. 2003). ABA also acts as an endogenous messenger, is generated as a secondary signaling molecule and is considered as an essential mediator of plant water balance regulation and osmotic stress tolerance (Negin and Moshelion 2016). Increased ABA levels in the plant enhance the response to water stress as it induces closure of the leaf stomata; which are microscopic pores involved in gas exchange (Daszkowska-Golec and Szarejko 2013). Stomatal closure results in a reduction of water loss through transpiration, thus decreasing the rate of photosynthesis (Mittler and Blumwald 2015). These changes have a short term effect and improve the water utilization efficiency of the plant. Different plants employing this ABA-regulated stress tolerance mechanism have been reported, including rice, barley, soybean, tomato, cotton and alfalfa (Sah et al. 2016). This ABA reaction is reversible; when the level of ABA drops, the stomata re-opens once water is available again (Tuteja 2007).

It is well established that the regulation of osmotic stress-response gene expression is maintained by two ABA pathways; the first is an ABA-dependent pathway in which ABA-Responsive Element (ABRE), ABRE-Binding (AREB) proteins or ABRE-Binding Factors (ABFs) as well as cis-acting elements are the major players. Furthermore, members of the plant Sucrose Non-fermenting 1-Related Kinase 2 (SnRK2) family of plant-specific serine/threonine kinases regulate the ABA-dependent pathway through phosphorylation of different proteins and transcription factors like AREB/ABFs which modulate the expression of stress responsive genes (Kulik et al. 2011; Yoshida et al. 2014). The second pathway involved in osmotic stress-response gene expression is conversely ABA-independent, where Dehydration-Responsive Element/C-Repeat (DRE/CRT), DRE−/CRT-Binding protein 2 (DREB2) transcription factors and cis-elements are the major players (Yoshida et al. 2014).

Another important player in regulating gene expression under abiotic stress is the Heat Shock Proteins (HSPs) family, which play an important role in thermotolerance reactions in plant cells. HSPs are suggested to act as molecular chaperones in protein quality control. Several HSPs encoded by Heat Shock Genes (HSGs) have been reported to be upregulated by heat stress (Hasanuzzaman et al. 2013). Several studies have demonstrated the involvement of HSPs in thermotolerance, however, a complete picture of the exact function and mechanisms utilized by these HSPs is still lacking (Ohama et al. 2016). Interestingly, HSPs have been shown to be involved in salt and drought stress signaling pathways as well (Jacob et al. 2016). In contrast, it has been reported that the DREB2A transcription factor plays a major role in the regulation and expression of drought stress responsive genes, and interestingly, the heat stress responsive genes as well (Yoshida et al. 2008).

4.3.1 DNA Microarray

DNA microarray analysis has been used extensively during the last two decades in plant gene expression studies. For instance, Arabidopsis microarray analysis identified 306 cold-responsive genes, belonging to multiple cold regulatory pathways (Fowler and Thomashow 2002). Rice (*Oryza sativa*) cDNA microarray has been used to identify salt, drought, cold and abscisic acid (ABA) inducible genes in rice (Rabbani et al. 2003). The review by Rensink and Buell discusses different resources for available plant microarrays (Rensink and Buell 2005). Additionally, a number of abiotic stress-inducible microRNAs (miRNAs) have been identified using microarray-based analysis suggesting their role in gene expression regulation under abiotic stresses (Liu et al. 2008). Kidokoro et al. (2015) reported the identification of 14 DREB1-type transcription factors in the Soybean (*Glycine max*) genome that have been shown to be induced by drought, salt, heat and cold abiotic stresses (Kidokoro et al. 2015). Arabidopsis plants expressing Soybean DREB1s have been analyzed using the Arabidopsis 3 Oligo Microarray (Agilent Technologies, http://www.agilent.com) and the microarray database at Genevestigator (https://www.genevestigator.com/gv/html.jsp), with many genes being upregulated by different abiotic stresses (Kidokoro et al. 2015). Table 4.1 shows examples of different databases, tools and recourses for plant genomics, proteomics, transcriptomics and metabolomics.

Table 4.1 Some online databases, tools and resources useful for plant omics studies

Name	Description	Link
Expression Atlas	Provides information on gene expression patterns under different biological conditions. Many plants are included	http://www.ebi.ac.uk/gxa/home
PlantGDB	Tools and resources for plant genomics	http://www.plantgdb.org/
Omictools	Search engin for biological data analysis. Many plants are included. Containing around 13411 omics tools	https://omictools.com/
Phytozome	A comparative platform for green plant genomics	https://phytozome.jgi.doe.gov/pz/
PLEXdb	Plant Expression Database. Gene expression resource for plants	http://www.plexdb.org
PlaNet	Platform of web-tools dedicated to visualization and analysis of plant co-function networks	http://aranet.mpimp-golm.mpg.de/
Gramene	Curated, open-source, integrated data resource for comparative functional genomics in crops and model plant species	http://www.gramene.org/
Ensembl Plants	Resource for plant genomics. Including browsers for Arabidopsis, rice, tomato, corn, and other species	http://plants.ensembl.org/index.html
PGSB	Providing data and information resource for individual plant species. A platform for integrative and comparative plant genome research	http://pgsb.helmholtz-muenchen.de/plant/athal/
GabiPD	A plant integrative omics database. Search for comprehensive and extensive information on various plant genomes	http://www.gabipd.org/
STRING	Database of known and predicted protein-protein interactions	http://string-db.org/
Araport	Arabidopsis Information Portal. Offers gene and protein reports with orthology, expression, interactions and the latest annotation, plus analysis tools for Arabidopsis	https://www.araport.org/
TAIR	The Arabidopsis Information Resource. A database of genetic and molecular biology data for Arabidopsis plant	http://www.arabidopsis.org/

4.3.2   Omics Technologies: the Recent Emerging Tools

More recently, with the aid of many plant omics databases, tools, and recourses (Table 4.1), omics technologies have been employed to identify target genes and the gene expression network associated with abiotic stress (Fig. 4.1). For example, transcriptomics analysis using bioinformatics and computational approaches have identified 32 WRKY transcription factors in Broomcorn millet (*Panicum miliaceum L.*), out of which 22 were differentially regulated under one or more of salt, drought, cold and heat stresses and considered as abiotic stress responsive genes (Ke et al. 2016). Transcriptomics analysis using the RNA-Seq approach in Arabidopsis plants exposed to a combination of salinity and heat stresses revealed several specific transcripts associated with the ABA pathway that responded to the stress combination but not either of them individually (Suzuki et al. 2016). Different potential functional and regulatory genes associated with the abiotic stress signalling pathway in rice have been identified through deep transcriptome sequencing using the Massively Parallel Signature Sequencing (MPSS) and Sequencing-By-Synthesis (SBS) technologies (Venu et al. 2013). A recent transcriptome profiling method has also been proposed to identify target genes and gene expression networks regulated by transcription factors (TFs). This method is called Infiltration-RNAseq and is a novel and fast technique coupling RNA-seq with TFs agroinfiltration (Bond et al. 2016).

4.4   Forward Genetics Strategies

Forward genetics in plants simply involves the identification of a gene responsible for a specific phenotype and linking the gene sequence with its function of displaying that specific phenotype. The concept of "forward genetics" that we know now is the same concept of "classical genetics" supported by modern and advanced technologies (Beutler et al. 2003). Mendel used the classical approach of "phenotype to gene" to study the natural variations in plant phenotypes. However, we now have technologies and tools such as mutagenesis and t-DNA approaches that help us to generate phenotypes (Fig. 4.2).

4.4.1   Mutagenesis

The induced mutation strategy, which was discovered in the early twentieth century, is considered one of the most powerful strategies for forward genetics. Mutations can be induced in plants through; (a) physical methods including; (i) ionizing radiation such as cosmic, gamma and X-rays, and (ii) ionizing particles such as alpha and beta particles as well as neutrons (Mba et al. 2010); (b) chemical methods including; (i) nitrous acid, (ii) base analogues such as 5-bromodeoxyuridine and 2-aminopurine (2AP), (iii) alkylating agents such as ethylmethane sulfonate (EMS),

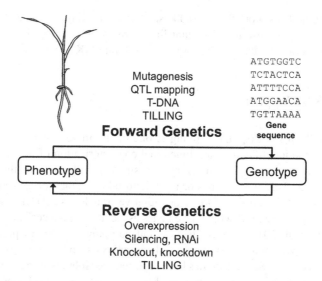

Fig. 4.2 Forward and reverse genetics in plants. Forward genetics focuses on identifying the gene causing a specific phenotype by using different strategies such as mutagenesis, QTL mapping, T-DNA etc. On the other hand, reverse genetics focuses on elucidating the gene function of a known sequence by using different strategies such as overexpression, RNAi, VIGS, etc

ethyl-2-chloroethyl sulphide, 2-chloroethyl-dimethyl amine and ethylene oxide (Mba et al. 2010). Several reports demonstrated the use of these physical and chemical methods to induce mutations in plants, and are discussed in detail in (Mba 2013). More recently, TILLING (Targeting Induced Local Lesions in Genomes) has been developed as a tool in forward and reverse genetics to induce random mutations through the genome (Jankowicz-Cieslak et al. 2017) by taking advantage of sequenced genomes and high throughput analysis in plant genomics. The general methodology for developing a TILLING in plants starts with acquiring a mutated population (M0 and M1), for example with a chemical method, followed by PCR amplification of the target sequences (Kurowska et al. 2011). Consequently, mutations in the M2 population can be detected through Denaturing High-Performance Liquid Chromatography (DHPLC), enzymatic mismatch cleavage, or sequencing strategies. Finally, the M3 and M4 populations are then analyzed phenotypically (Kurowska et al. 2011).

4.4.2   T-DNA

Insertional mutagenesis using transfer DNA (T-DNA) is a powerful approach for the functional genomics analysis of plant genes. The T-DNA inserted randomly into the target plant genome disrupts the gene expression of the gene in which it is inserted, and is also used as a marker for mutant identification (Krysan et al. 1999). Arabidopsis indexed T-DNA inserts can be found at (http://signal.salk.edu/cgi-bin/tdnaexpress)

which accommodates more than 250,000 sequences (Daxinger et al. 2008). Beside Arabidopsis, T-DNA insertional mutant lines have been developed in other plant species such as rice, maize, barley, sorghum and tomato (Kuromori et al. 2009).

4.4.3 QTL Mapping

Quantitative Trait Loci (QTL) mapping is one of the important forward genetics strategies for the identification and functional characterization of potential regulatory genes or allelic variants under abiotic stresses. Tolerance to these abiotic stresses is controlled by polygenes or multifactorial traits that are generally called "quantitative traits", and therefore, a quantitative trait locus (QTL) describes the regions in the plant genome which have genes associated with a certain quantitative trait (Sehgal et al. 2016). QTL detection and mapping analysis in plants is performed using two common methods; linkage analysis and association mapping (Semagn et al. 2010). Several studies have reported the isolation of numerous QTLs that are associated with abiotic stresses. For example, a rice QTL designated as SKC1 has been mapped and found to play a major role in salt stress tolerance by K+/Na+homeostasis regulation in rice (Ren et al. 2005). A recent study by Fan et al. 2015 employed a QTL mapping strategy for the identification of drought and salt tolerance QTLs in the barley population via double haploid lines generated by a cross between a drought and salinity tolerance variety (TX9425) and a sensitive variety (Franklin) (Fan et al. 2015).

4.5 Reverse Genetics Strategies

Advances in genome sequencing technologies and bioinformatics tools have rapidly grown in recent years to allow faster sequencing of a multitude of plant genomes. However, the functional characterization of this sequencing and an understanding of the pathways controlling the associated genetic mechanisms remains elusive. Reverse genetics is a term used to describe the elucidation of the unknown function of a known gene sequence. Different reverse genetic strategies have been employed to understand the function of sequenced genes by a gain of function (such as gene overexpression) or a loss of function (such as RNA Interference or RNAi, Virus-Induced Gene Silencing or VIGS, TILLING etc.) (Fig. 4.2).

4.5.1 Overexpression

A straightforward strategy for understanding the function of a plant gene associated with abiotic stresses is to overexpress it in its wild type plant species, followed by phenotypic, biochemical, physiological etc. analysis to compare the overexpressed transgenic plants with wild type plants under the tested abiotic stress (e.g. drought, salinity and heat).

Countless studies have applied such an approach for the functional characterization of several plant genes and transcription factors associated with abiotic stress tolerance and for the development of abiotic stress tolerance in plants. For example, the extensive review by Puranik et al. (2012) summarized the functional characterization of several NAC members through overexpression approaches and emphasized a vital role for NAC transcription factors in plant abiotic stress responses. NAC transcription factors have a NAC domain and consist of a big superfamily in plants with more than 100 members (Puranik et al. 2012). NAC (NAM, AFAT, and CUC) stands for No Apical Meristem (NAM), Arabidopsis Transcription Activation Factor (ATAF) and Cup-shaped Cotyledon (CUC) (Puranik et al. 2012). A recent study by Hong et al. 2016 identified a novel NAC gene, ONAC022, in rice. Transgenic rice lines overexpressing ONAC022 exhibited increased drought and salt tolerance (Hong et al. 2016). Evidence of the important role of WRKY transcription factors has also been demonstrated through different overexpression studies (Rushton et al. 2010). Wu et al. (2009) reported that transgenic rice plants overexpressing the rice WRKY11 gene encoding for the WRKY11 transcription factor under the HSP101 rice promoter exhibited tolerance to heat and drought (Wu et al. 2009). In contrast, during Arabidopsis seed germination, overexpression of AtWRKY30 in Arabidopsis under the Cauliflower Mosaic Virus (CaMV) 35S promoter increased tolerance to oxidative and salinity stresses (Scarpeci et al. 2013). Overexpression of rice aquaporin genes PIP2;4 and PIP2;7 in Arabidopsis exhibited a more tolerant phenotype under arsenite (Mosa 2012) and boron (Kumar K et al. 2014) toxicity. Additionally, transgenic Arabidopsis plants overexpressing OsPIP2;6 also showed increased tolerance to arsenite (Mosa et al. 2012) and boron (Mosa et al. 2016). The above-mentioned studies suggested that these rice PIP aquaporin members facilitate the bidirectional transport of metalloids such as arsenic in the form of arsenite and boron in the form of boric acid. Members of the Ethylene Response Factor (ERF) family have demonstrated important regulatory roles in response to salt and drought stresses (Mizoi et al. 2012). Transgenic tobacco plants overexpressing soybean ERF3 showed enhanced tolerance to drought and salt stresses and accumulated more free proline and soluble carbohydrates in comparison to wild-type plants (Zhang et al. 2009). Similarly, overexpression of wheat ERF3 in wheat enhanced its tolerance to drought and salt stresses (Rong et al. 2014).

4.5.2 RNAi

RNA interference (RNAi) is one of the most powerful approaches to induce gene silencing in plants. RNA silencing pathways can be activated by introducing a construct harboring a single or double-stranded RNA complementary to the gene of interest that can degrade the transcript of the gene of interest that is to be silenced (Agrawal et al. 2003; Gilchrist and Haughn 2010; Younis et al. 2014). The general strategy of plant genetic engineering via the use of RNAi includes four steps, starting with the identification of target gene pathways by performing the bioinformatics analysis of the target genome, transcriptome, proteome, and metabolome. The second step is selecting a suitable vector along with its appropriate promoter to prepare the RNAi and to enable the screening of RNAi constructs by the addition of selectable markers into the vector. The third step includes the delivery of RNAi into the plant and screening the transformed plant after performing a tissue culture of the transgenic plant. The final step is accomplished by evaluating the transgenic RNAi plants by comparing it with wild type plants.

4.5.3   VIGS

Virus-Induced Gene Silencing (VIGS) is a fast technique that is employed in the field of plant functional genomics to reveal gene functions and is applied to both forward and reverse genetics. Gene silencing by the VIGS method can be achieved by activating the Small Interfering RNA (siRNA) silencing pathways via transfecting the target plant with the DNA copy of an RNA plant virus harboring a fragment of 200–1300 bp from the plant target gene. The double-stranded RNA resulting from viral replication inside the target plant cells is degraded by the Dicer-like enzyme into siRNA (Hayward et al. 2011; Gilchrist and Haughn 2013). Several studies have utilized VIGS tools for the functional characterization of different plant genes in many plant species. For instance, recombinant Barley Stripe Mosaic Virus (BSMV) harboring 222 bp of wheat ERF3 was used to silence the ERF3 gene in wheat. The wheat ERF3 silenced plants exhibited sensitivity phenotypes to salt and drought stresses, suggesting its importance in responding to these conditions (Rong et al. 2014).

4.5.4   Yeast Complementation Assays

The yeast functional complementation assay is a fast, simple and efficient approach for assessing the function of plant genes. The target plant gene to be tested is cloned into a yeast expression vector and then expressed in a yeast mutant lacking the homologous gene in the yeast system in which its function is known. The expressed gene is evaluated for its ability to complement the function of the mutated gene in yeast (Fig. 4.3). Numerous studies have applied the yeast complementation assay strategy for the functional characterization of different plant genes. For instance, yeast complementation analysis of the heavy metal hyperaccumulator plant species *Thlaspi caerulescens* identified different P-type ATPase and metallothionein genes that exhibited cadmium tolerance in yeast (Papoyan and Kochian 2004). Heterologous expression of a vacuolar Na+/H+ antiporter gene NHX1 isolated from cowpea (*Vigna unguiculata L.*) in the AXT3 yeast mutant strain lacking the plasma

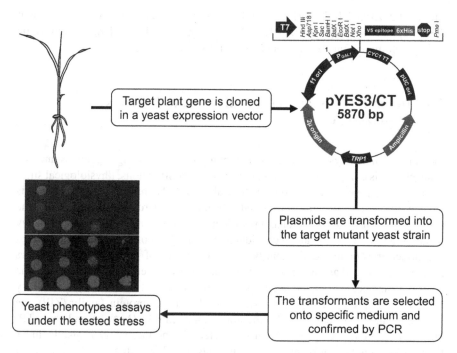

Fig. 4.3 Simplified diagram showing the steps involved in a yeast functional complementation assay. The target gene isolated from the target plant species is cloned on a yeast specific expression vector such as pYES3 for subsequent transformation into the target yeast mutant strain. The transformants are then selected and confirmed by PCR. Yeast culture of the cells expressing the plant gene is grown on the appropriate medium before yeast phenotypes assays are performed under the tested abiotic stress. The yeast picture used in this figure is taken from Mosa (2012)

membrane antiporters (Na + effluxer, and Na + vacuolar) showed that cowpea NHX1 complemented for the loss of yeast NHX1 under NaCl/KCl stress (Mishra et al. 2015). Kumar K et al. (2014) reported that expression of rice aquaporin transporters PIP2;4 and PIP2;7 in the yeast HD9 strain lacking the aquaporin transporter fps1, efflux pump acr3, and vacuolar transporter ycf enhanced yeast sensitivity under boron treatment and increased boron uptake in yeast cells as a result of the ability of these rice aquaporin genes to complement the function of the mutated aquaporin yeast strain (Kumar K et al. 2014). More recently, rice aquaporins PIP1;3 and PIP2;6 expressed in the HD9 yeast strain also resulted in increased boron sensitivity and accumulated a higher level of boron in comparison with control yeast cells expressing the empty vector (Mosa et al. 2016). Eleven members of the Arabidopsis ZIP micronutrient transporter gene family have been heterologously expressed in different yeast mutants lacking Fe uptake (fet3/fet4Δ), Cu uptake (ctr1/ctr3Δ), Zn uptake (zrt1/zrt2Δ) or Mn uptake (smf1Δ) (Milner et al. 2013). Arabidopsis ZIP7 exhibited its ability to complement the Fe uptake mutant, whereas all eleven ZIP members did not show any ability to complement the Cu uptake mutant. Interestingly,

ZIP1, ZIP2, ZIP3, ZIP7, ZIP11, and ZIP12 demonstrated the ability to complement the Zn uptake mutant, while ZIP1, ZIP2, ZIP5, ZIP6, ZIP7, and ZIP9 showed the ability to complement the Mn uptake mutant (Milner et al. 2013).

4.6 Ionomics as a Powerful Tool for Abiotic Stress Exploration

Ionomics is the analysis of the elemental composition of living organisms and any changes in this composition as a result of genetic modifications, physiological stimuli and developmental conditions (Salt et al. 2008). Therefore, ionomics analysis has the potential to provide substantial information regarding the genes and gene networking that controls the elemental composition under abiotic stresses. Heavy metals and toxicity have gained considerable attention as one of the abiotic stresses that affect crop quality and quantity. Several families of plant metal transporters have been identified and are discussed in detail in (Komal et al. 2015), including the natural resistance-associated macrophage protein family (NRAMP), the heavy metal ATPase protein family (HMA), the ZRT-IRT-like protein family (ZIP), the multidrug and toxic compound extrusion protein family (MATE) and the ATP-binding cassette (ABC) protein family etc. Hence, ionomics analysis also includes the possible function of plant mineral nutrients in reducing the toxicity caused by heavy metals (Singh et al. 2015). For example, cadmium toxicity was reduced by sulphur application in mustard (*Brassica campestris L.*) plants by increasing ascorbate and glutathione concentrations which enhanced photosynthesis and plant growth (Anjum et al. 2008). Moreover, cadmium alleviation has also been reported in rice plants through the application of calcium and silica, which suppressed oxidative damage caused by cadmium toxicity and decreased the level of reactive oxygen species (ROS) (Srivastava et al. 2015).

The elemental composition of the plant ionome may be changed as a result of nutrient concentration changes in soil or alterations in ion transporter proteins (Salt et al. 2008; Singh et al. 2015). Transporters are protein complexes which span the membranes that surround the plant cells; they transport molecules and ions from the soil to inside the plant cells, from one cell to another, and between cells and vessels such as phloem and xylem (Sondergaard et al. 2004). Manipulating the transporter proteins localized within cell membranes have a great impact in making crops more tolerant to stresses and more efficient in the use of nutrients, in addition to controlling drought tolerance and transporting sucrose to where it is needed (Schroeder et al. 2013). Transporters also allow the plant to withstand adverse environments with increased salinity and acidity of soils. The aquaporin superfamily is an important example of transporter proteins associated with different abiotic stresses. Plant aquaporins are channel proteins that facilitate the transport of water and other small neutral molecules across the plant cell membrane (Johansson et al. 2000). Plant aquaporins have been classified into five major classes according to their sequence similarities and localization; the Nodulin26-like Intrinsic Proteins (NIPs), the

Plasma membrane Intrinsic Proteins (PIPs), the Small basic Intrinsic Proteins (SIPs), the Tonoplast Intrinsic Proteins (TIPs) and the uncategorized (X) Intrinsic Proteins (XIPs) (Vialaret et al. 2014). In addition to their role as transporters for water, urea, glycerol, metalloids, NH_3, CO_2 and reactive oxygen species (ROS) (Vialaret et al. 2014; Afzal et al. 2016), several plant aquaporins have also been reported for their involvement in plant abiotic stress tolerance through their transport function (Afzal et al. 2016). For example, transgenic Arabidopsis plants overexpressing a wheat NIP gene accumulated higher levels of K+, Ca²+ and proline and lower levels of Na+, suggesting its involvement in plant salt tolerance mechanisms and signaling pathways (Gao et al. 2010).

A successful approach to understanding salt tolerance mechanisms in plants is to control the ion transporters involved in influx, efflux, compartmentation and translocation of toxic ions, which exhibits their role in salt stress tolerance. Different ion transporters associated with salt tolerance mechanisms have been identified in plants which are discussed in detail in (Kumar and Mosa 2015); including Na+/H+ Antiporters (NHX), Cation Antiporters (CAX), High-Affinity Potassium Transporters (HKTs), Salt Overlay Sensitive (SOS1) transporters and Nonselective Cation Channels (NSCCs) transporters.

4.7 Metabolomics to Discover Abiotic Stress Related Phytochemicals

Metabolomics analysis simply involves the identification and quantification of low molecular weight metabolites in an organism, which could be a specific organ, tissue, cell, or alternatively at a specific developmental stage or under specific environmental conditions (Allakhverdiev et al. 2008; Arbona et al. 2013). The plant metabolome is considered as the link between genotype and phenotype (Arbona et al. 2009), which makes it a powerful tool in functional genomics to analyze metabolites under certain environmental factors (Roessner et al. 2001) such as abiotic stresses. Two approaches have been used to study plant metabolomics under different abiotic stresses; untargeted metabolomics analysis which focuses on metabolite profiling that assesses as many metabolites as possible under the tested abiotic stress (De Vos et al. 2007), while targeted metabolomics analysis aims at the identification and quantification of specific metabolites. Figure 4.4 represents the workflow of plant metabolomics analysis to identify abiotic stress-related metabolites.

Examples of plant metabolites that are associated with abiotic stresses are compounds belonging to the osmolytes group (Gupta and Huang 2014). Osmolytes (which are naturally produced organic compounds) such as amino acids (e.g. free proline), quaternary ammonium compounds (e.g. glycinebetaine), soluble sugars (e.g. mannitol, fructans, sorbitol and trehalose) and sugar alcohols (e.g. polyols) (Khan et al. 2010) accumulate in plant cells in response to abiotic stresses. High levels of osmolytes in plant cells contribute to adjusting the intracellular osmotic potential,

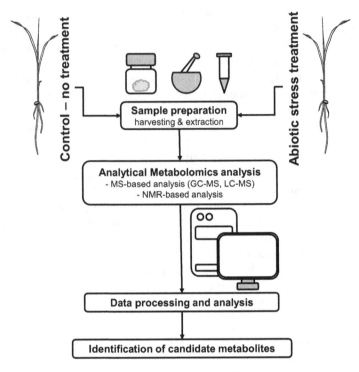

Fig. 4.4 Workflow of plant metabolomics analysis to identify abiotic stress-related metabolites. Plant samples from control and treated plants are harvested and extracted using the appropriate extraction buffer before analysis through NMR-based or MS-based instruments including LC-MS and GC-MS. The resulting data is then processed and analyzed using different bioinformatics tools to identify the candidate abiotic stress-related metabolites

maintaining the structure of proteins, reduction in the size of the cell environment and preserving the higher energy status in order to protect cell proteins (Gupta and Huang 2014). Over-expression of the bacterial choline oxidase gene (codA) in Arabidopsis plants leads to an increase in the biosynthesis of glycinebetaine metabolites through the catalyzation of choline into glycinebetaine; this results in increasing the tolerance to high-temperature stress in transgenic Arabidopsis in comparison to wild-type Arabidopsis (Alia et al. 1998). Furthermore, increased tolerance to drought and salt stress was reported in transgenic poplar plants (*Populus alba × Populus glandulosa*) expressing the codA gene (Ke et al. 2016). Spinach BADH protein overexpressed in tobacco plants has been shown to increase the levels of glycinebetaine and therefore provided increased high temperature tolerance as well as the preservation of CO_2 and rubisco activity assimilation rates under heat stress (Yang et al. 2005). Trehalose metabolites provide osmoprotection against the reverse effect of desiccation from drought, decreased temperature or salt stresses in the environment (Iordachescu and Imai 2011). High accumulation of glycine betaine and trehalose metabolites have been reported in tomato plants exposed to a combination of salinity and heat stress, suggesting a potential role for these metabolites in protective tolerance mechanisms against this stress combination (Rivero et al. 2014).

References

Afzal Z, Howton T, Sun Y, Mukhtar M (2016) The Roles of Aquaporins in Plant Stress Responses. J Dev Biol 4:9. doi: 10.3390/jdb4010009

Agrawal N, Dasaradhi PVN, Mohmmed A, et al (2003) RNA interference: biology, mechanism, and applications. Microbiol Mol Biol Rev 67:657–685. doi: 10.1128/MMBR.67.4.657

Alia, Hayashi H, Sakamoto A, Murata N (1998) Enhancement of the tolerance of Arabidopsis to high temperatures by genetic engineering of the synthesis of glycinebetaine. Plant J 16:155–161. doi: 10.1046/j.1365-313X.1998.00284.x

Allakhverdiev SI, Kreslavski VD, Klimov V V, et al (2008) Heat stress: An overview of molecular responses in photosynthesis. Photosynth Res 98:541–550. doi: 10.1007/s11120-008-9331-0

Anjum NA, Umar S, Ahmad A, et al (2008) Sulphur protects mustard (Brassica campestris L.) from cadmium toxicity by improving leaf ascorbate and glutathione: Sulphur protects mustard from cadmium toxicity. Plant Growth Regul 54:271–279. doi: 10.1007/s10725-007-9251-6

Arbona V, Iglesias DJ, Talón M, Gómez-Cadenas A (2009) Plant phenotype demarcation using nontargeted LC-MS and GC-MS metabolite profiling. J Agric Food Chem 57:7338–7347. doi: 10.1021/jf9009137

Arbona V, Manzi M, de Ollas C, Gómez-Cadenas A (2013) Metabolomics as a tool to investigate abiotic stress tolerance in plants. Int J Mol Sci 14:4885–4911. doi: 10.3390/ijms14034885

Beutler B, Du X, Hoebe K (2003) From phenomenon to phenotype and from phenotype to gene: forward genetics and the problem of sepsis. J Infect Dis 187 Suppl:S321–6. doi: 10.1086/374757

Bond DM, Albert NW, Lee RH, et al (2016) Infiltration-RNAseq: transcriptome profiling of Agrobacterium-mediated infiltration of transcription factors to discover gene function and expression networks in plants. Plant Methods 12:41. doi: 10.1186/s13007-016-0141-7

Daszkowska-Golec A, Szarejko I (2013) Open or close the gate - stomata action under the control of phytohormones in drought stress conditions. Front Plant Sci 4:138. doi: 10.3389/fpls.2013.00138

Daxinger L, Hunter B, Sheikh M, et al (2008) Unexpected silencing effects from T-DNA tags in Arabidopsis. Trends Plant Sci 13:4–6. doi: 10.1016/j.tplants.2007.10.007

De Vos RCH, Moco S, Lommen A, et al (2007) Untargeted large-scale plant metabolomics using liquid chromatography coupled to mass spectrometry. Nat Protoc 2:778–791. doi: 10.1038/nprot.2007.95

Fan Y, Shabala S, Ma Y, et al (2015) Using QTL mapping to investigate the relationships between abiotic stress tolerance (drought and salinity) and agronomic and physiological traits. BMC Genomics 16:43. doi: 10.1186/s12864-015-1243-8

Fowler S, Thomashow MF (2002) Arabidopsis transcriptome profiling indicates that multiple regulatory pathways are activated during cold acclimation in addition to the CBF cold response pathway. Plant Cell 14:1675–1690. doi: 10.1105/tpc.003483.Toward

Furbank RT, Tester M (2011) Phenomics—technologies to relieve the phenotyping bottleneck. Trends Plant Sci 16:635–644. doi: 10.1016/j.tplants.2011.09.005

Gao Z, He X, Zhao B, et al (2010) Overexpressing a putative aquaporin gene from wheat, TaNIP, enhances salt tolerance in transgenic arabidopsis. Plant Cell Physiol 51:767–775. doi: 10.1093/pcp/pcq036

Gilchrist E, Haughn G (2010) Reverse genetics techniques: engineering loss and gain of gene function in plants. Brief Funct Genomics 9(2):103–110. doi: https://doi.org/10.1093/bfgp/elp059. https://academic.oup.com/bfg/article/9/2/103/216226/Reverse-genetics-techniques-engineering-loss-and

Gilchrist E, Haughn G (2013) Gene Identification: Reverse Genetics. Diagnostics Plant Breed. Springer, pp 61–89

Gupta B, Huang B (2014) Mechanism of salinity tolerance in plants: Physiological, biochemical, and molecular characterization. Int J Genomics. doi: 10.1155/2014/701596

Hasanuzzaman M, Nahar K, Alam MM, et al (2013) Physiological, biochemical, and molecular mechanisms of heat stress tolerance in plants. Int J Mol Sci 14:9643–9684. doi: 10.3390/ijms14059643

Hayward A, Padmanabhan M, Dinesh-Kumar SP (2011) Virus-induced gene silencing in nicotiana benthamiana and other plant species. Methods Mol Biol 678:55–63. doi:10.1007/978-1-60761-682-5_5

Hong Y, Zhang H, Huang L, et al (2016) Overexpression of a Stress-Responsive NAC Transcription Factor Gene ONAC022 Improves Drought and Salt Tolerance in Rice. Front Plant Sci 7:1–19. doi: 10.3389/fpls.2016.00004

Iordachescu M, Imai R (2011) Trehalose and Abiotic Stress in Biological Systems. Abiotic Stress Plants - Mech. Adapt. pp 215–234

Jacob P, Hirt H, Bendahmane A (2016) The heat shock protein/chaperone network and multiple stress resistance. Plant Biotechnol. J. 15:405–414

Jankowicz-Cieslak, J., Mba, C., & Till, B. J. (2017). Mutagenesis for Crop Breeding and Functional Genomics. In Biotechnologies for Plant Mutation Breeding (pp. 3-18). Springer International Publishing. doi: 10.1007/978-3-319-45021-6_1. https://link.springer.com/chapter/10.1007/978-3-319-45021-6_1

Johansson I, Karlsson M, Johanson U, et al (2000) The role of aquaporins in cellular and whole plant water balance. Biochim Biophys Acta - Biomembr 1465:324–342. doi: 10.1016/S0005-2736(00)00147-4

Ke Q, Wang Z, Ji CY, et al (2016) Transgenic poplar expressing codA exhibits enhanced growth and abiotic stress tolerance. Plant Physiol Biochem 100:75–84. doi: 10.1016/j.plaphy.2016.01.004

Khan SH, Ahmad N, Ahmad F, Kumar R (2010) Naturally occurring organic osmolytes: From cell physiology to disease prevention. IUBMB Life 62:891–895. doi: 10.1002/iub.406

Kidokoro S, Watanabe K, Ohori T, et al (2015) Soybean DREB1/CBF-type transcription factors function in heat and drought as well as cold stress-responsive gene expression. Plant J 81:505–518. doi: 10.1111/tpj.12746

Komal, T., Mustafa, M., Ali, Z., & Kazi, A. G. (2015). Heavy metal uptake and transport in plants. In Heavy metal contamination of soils (pp. 181–194). Volume 44 of the series Soil Biology. Springer International Publishing. https://link.springer.com/chapter/10.1007/978-3-319-14526-6_10

Krysan PJ, Young JC, Sussman MR (1999) T-DNA as an insertional mutagen in Arabidopsis. Plant Cell 11:2283–2290. doi:10.1105/tpc.11.12.2283

Kulik A, Wawer I, Krzywińska E, et al (2011) SnRK2 Protein Kinases—Key Regulators of Plant Response to Abiotic Stresses. Omi A J Integr Biol 15:859–872. doi: 10.1089/omi.2011.0091

Kumar, K., & Mosa, K. A. (2015). Ion Transporters: A Decisive Component of Salt Stress Tolerance in Plants. In Managing Salt Tolerance in Plants: Molecular and Genomic Perspectives (pp. 373–390). CRC Press. doi: 10.1201/b19246-21. http://www.crcnetbase.com/doi/abs/10.1201/b19246-21

Kumar K, Mosa KA, Chhikara S, et al (2014) Two rice plasma membrane intrinsic proteins, OsPIP2;4 and OsPIP2;7, are involved in transport and providing tolerance to boron toxicity. Planta 239:187–198. doi: 10.1007/s00425-013-1969-y

Kumar A, Kage U, Mosa K, Dhokane D (2014) Metabolomics: A Novel Tool to Bridge Phenome to Genome under Changing Climate to Ensure Food Security. Med Aromat Plants 3:e154. doi:10.4172/2167-0412.1000e154. https://www.omicsgroup.org/journals/metabolomics-a-novel-tool-to-bridge-phenome-to-genome-under-changingclimate-to-ensure-food-security-2167-0412.1000e154.php?aid=33497

Kuromori T, Takahashi S, Kondou Y, et al (2009) Phenome analysis in plant species using loss-of-function and gain-of-function mutants. Plant Cell Physiol 50:1215–1231. doi: 10.1093/pcp/pcp078

Kurowska M, Daszkowska-Golec A, Gruszka D, et al (2011) TILLING - a shortcut in functional genomics. J Appl Genet 52:371–390. doi: 10.1007/s13353-011-0061-1

Kushalappa AC, Gunnaiah R (2013) Metabolo-proteomics to discover plant biotic stress resistance genes. Trends Plant Sci 18:522–531. doi: 10.1016/j.tplants.2013.05.002

Liu H-H, Tian X, Li Y-J, et al (2008) Microarray-based analysis of stress-regulated microRNAs in Arabidopsis thaliana. RNA 14:836–843. doi: 10.1261/rna.895308

Mba C (2013) Induced Mutations Unleash the Potentials of Plant Genetic Resources for Food and Agriculture. Agronomy 3:200–231. doi: 10.3390/agronomy3010200

Mba, C., Afza, R., Bado, S., & Jain, S. M. (2010). Induced mutagenesis in plants using physical and chemical agents. Plant cell culture: essential methods, 20, 111-130. doi: 10.1002/9780470686522. ch7. http://onlinelibrary.wiley.com/doi/10.1002/9780470686522.ch7/summary

Mihaela Iordachescu and Ryozo Imai (2011). Trehalose and Abiotic Stress in Biological Systems, Abiotic Stress in Plants - Mechanisms and Adaptations, Arun Shanker (Ed.), InTech, doi: 10.5772/22208. https://www.intechopen.com/books/abiotic-stress-in-plants-mechanisms-and-adaptations/trehalose-and-abioticstress-in-biological-systems

Milner MJ, Seamon J, Craft E, Kochian L V (2013) Transport properties of members of the ZIP family in plants and their role in Zn and Mn homeostasis. J Exp Bot 64:369–381. doi: 10.1093/jxb/ers315

Mishra S, Alavilli H, Lee B, et al (2015) Cloning and characterization of a novel vacuolar Na+/H+ antiporter gene (VuNHX1) from drought hardy legume, cowpea for salt tolerance. Plant Cell Tissue Organ Cult 120:19–33.

Mittler R, Blumwald E (2015) The roles of ROS and ABA in systemic acquired acclimation. Plant Cell 27:64–70. doi: 10.1105/tpc.114.133090

Mizoi J, Shinozaki K, Yamaguchi-Shinozaki K (2012) AP2/ERF family transcription factors in plant abiotic stress responses. Biochim Biophys Acta - Gene Regul Mech 1819:86–96. doi: 10.1016/j.bbagrm.2011.08.004

Mosa, K. A. (2012). Functional characterization of members of plasma membrane intrinsic proteins subfamily and their involvement in metalloids transport in plants. Doctoral Dissertations Available from Proquest. Paper AAI3518265. http://scholarworks.umass.edu/dissertations/AAI3518265

Mosa KA, Kumar K, Chhikara S, et al (2012) Members of rice plasma membrane intrinsic proteins subfamily are involved in arsenite permeability and tolerance in plants. Transgenic Res 21:1265–1277. doi: 10.1007/s11248-012-9600-8

Mosa KA, Kumar K, Chhikara S, et al (2016) Enhanced Boron Tolerance in Plants Mediated by Bidirectional Transport Through Plasma Membrane Intrinsic Proteins. Sci Rep 6:21640. doi: 10.1038/srep21640

Nakashima K, Yamaguchi-Shinozaki K, Shinozaki K (2014) The transcriptional regulatory network in the drought response and its crosstalk in abiotic stress responses including drought, cold, and heat. Front Plant Sci 5:170. doi: 10.3389/fpls.2014.00170

Negin B, Moshelion M (2016) The evolution of the role of ABA in the regulation of water-use efficiency: From biochemical mechanisms to stomatal conductance. Plant Sci 251:82–89. doi: 10.1016/j.plantsci.2016.05.007

Ohama N, Sato H, Shinozaki K, Yamaguchi-Shinozaki K (2016) Transcriptional Regulatory Network of Plant Heat Stress Response. Trends Plant Sci doi: 10.1016/j.tplants.2016.08.015

Papoyan A, Kochian L V (2004) Identification of Thlaspi caerulescens Genes That May Be Involved in Heavy Metal Hyperaccumulation and Tolerance. Characterization of a Novel Heavy Metal Transporting ATPase 1. Plant Physiol 136:3814–3823. doi: 10.1104/pp.104.044503

Puranik S, Sahu PP, Srivastava PS, Prasad M (2012) NAC proteins: Regulation and role in stress tolerance. Trends Plant Sci 17:369–381. doi: 10.1016/j.tplants.2012.02.004

Rabbani MA, Maruyama K, Abe H, et al (2003) Monitoring expression profiles of rice genes under cold, drought, and high-salinity stresses and abscisic acid application using cDNA microarray and RNA gel-blot analyses. Plant Physiol 133:1755–1767. doi: 10.1104/pp.103.025742

Ray DK, Mueller ND, West PC, Foley JA (2013) Yield Trends Are Insufficient to Double Global Crop Production by 2050. PLoS One doi:10.1371/journal.pone.0066428

Ren Z-H, Gao J-P, Li L, et al (2005) A rice quantitative trait locus for salt tolerance encodes a sodium transporter. Nat Genet 37:1141–1146. doi: 10.1038/ng1643

Rensink WA, Buell CR (2005) Microarray expression profiling resources for plant genomics. Trends Plant Sci 10:603–609. doi: 10.1016/j.tplants.2005.10.003

Rivero RM, Mestre TC, Mittler R, et al (2014) The combined effect of salinity and heat reveals a specific physiological, biochemical and molecular response in tomato plants. Plant Cell Environ 37:1059–1073. doi: 10.1111/pce.12199

Roessner U, Luedemann A, Brust D, et al (2001) Metabolic profiling allows comprehensive phenotyping of genetically or environmentally modified plant systems. Plant Cell 13:11–29. doi: 10.1105/tpc.13.1.11

Rong W, Qi L, Wang A, et al (2014) The ERF transcription factor TaERF3 promotes tolerance to salt and drought stresses in wheat. Plant Biotechnol J 12:468–479. doi: 10.1111/pbi.12153

Rushton PJ, Somssich IE, Ringler P, Shen QJ (2010) WRKY transcription factors. Trends Plant Sci 15:247–258. doi: 10.1016/j.tplants.2010.02.006

Sah SK, Reddy KR, Li J (2016) Abscisic Acid and Abiotic Stress Tolerance in Crop Plants. Front Plant Sci 7:571. doi: 10.3389/fpls.2016.00571

Salt DE, Baxter I, Lahner B (2008) Ionomics and the Study of the Plant Ionome. Annu Rev Plant Biol 59:709–733. doi: 10.1146/annurev.arplant.59.032607.092942

Scarpeci TE, Zanor MI, Mueller-Roeber B, Valle EM (2013) Overexpression of AtWRKY30 enhances abiotic stress tolerance during early growth stages in Arabidopsis thaliana. Plant Mol Biol 83:265–277. doi: 10.1007/s11103-013-0090-8

Schroeder JI, Delhaize E, Frommer WB, et al (2013) Using membrane transporters to improve crops for sustainable food production. Nature 497:60–66. doi: 10.1038/nature11909

Sehgal, D., Singh, R., & Rajpal, V. R. (2016). Quantitative Trait Loci Mapping in Plants: Concepts and Approaches. In Molecular Breeding for Sustainable Crop Improvement (pp. 31–59). Volume 11 of the series Sustainable Development and Biodiversity. Springer International Publishing. https://link.springer.com/chapter/10.1007/978-3-319-27090-6_2

Semagn K, Bjornstad Å, Xu Y (2010) The genetic dissection of quantitative traits in crops. Electron J Biotechnol doi: 10.2225/vol13-issue5-fulltext-21

Shinozaki K, Yamaguchi-Shinozaki K, Seki M (2003) Regulatory network of gene expression in the drought and cold stress responses. Curr Opin Plant Biol 6:410–417. doi: S136952660300092X [pii]

Singh S, Parihar P, Singh R, et al (2015) Heavy metal tolerance in plants: Role of transcriptomics, proteomics, metabolomics and ionomics. Front Plant Sci doi: 10.3389/fpls.2015.01143

Sondergaard TE, Schulz A, Palmgren MG (2004) Energization of transport processes in plants. roles of the plasma membrane H+−ATPase. Plant Physiol 136:2475–2482. doi: 10.1104/pp.104.048231

Srivastava RK, Pandey P, Rajpoot R, et al (2015) Exogenous application of calcium and silica alleviates cadmium toxicity by suppressing oxidative damage in rice seedlings. Protoplasma 252:959–975.

Suzuki N, Bassil E, Hamilton JS, et al (2016) ABA is required for plant acclimation to a combination of salt and heat stress. PLoS One doi: 10.1371/journal.pone.0147625

Türktaş M, Kurtoglu KY, Dorado G, et al (2015) Sequencing of plant genomes—A review. Turkish J Agric For 39:361–376. doi: 10.3906/tar-1409-93

Tuteja N (2007) Abscisic Acid and abiotic stress signaling. Plant Signal Behav 2:135–138. doi: 10.1111/j.1365-3040.2011.02426.x

Venu RC, Sreerekha M V, Sheshu Madhav M, et al (2013) Deep transcriptome sequencing reveals the expression of key functional and regulatory genes involved in the abiotic stress signaling pathways in rice. J Plant Biol 56:216–231. doi: 10.1007/s12374-013-0075-9

Vialaret J, Di Pietro M, Hem S, et al (2014) Phosphorylation dynamics of membrane proteins from Arabidopsis roots submitted to salt stress. Proteomics 14:1058–1070. doi: 10.1002/pmic.201300443

Wu X, Shiroto Y, Kishitani S, et al (2009) Enhanced heat and drought tolerance in transgenic rice seedlings overexpressing OsWRKY11 under the control of HSP101 promoter. Plant Cell Rep 28:21–30. doi: 10.1007/s00299-008-0614-x

Yamasaki K, Kigawa T, Seki M, et al (2013) DNA-binding domains of plant-specific transcription factors: Structure, function, and evolution. Trends Plant Sci 18:267–276. doi: 10.1016/j.tplants.2012.09.001

Yang X, Liang Z, Lu C (2005) Genetic engineering of the biosynthesis of glycinebetaine enhances photosynthesis against high temperature stress in transgenic tobacco plants. Plant Physiol 138:2299–2309. doi: 10.1104/pp.105.063164

Yoshida T, Mogami J, Yamaguchi-Shinozaki K (2014) ABA-dependent and ABA-independent signaling in response to osmotic stress in plants. Curr Opin Plant Biol 21:133–139. doi: 10.1016/j.pbi.2014.07.009

Yoshida T, Sakuma Y, Todaka D, et al (2008) Functional analysis of an Arabidopsis heat-shock transcription factor HsfA3 in the transcriptional cascade downstream of the DREB2A stress-regulatory system. Biochem Biophys Res Commun 368:515–521. doi: 10.1016/j.bbrc.2008.01.134

Younis A, Siddique MI, Kim CK, Lim KB (2014) RNA interference (RNAi) induced gene silencing: A promising approach of hi-tech plant breeding. Int J Biol Sci 10:1150–1158. doi: 10.7150/ijbs.10452

Zhang G, Chen M, Li L, et al (2009) Overexpression of the soybean GmERF3 gene, an AP2/ERF type transcription factor for increased tolerances to salt, drought, and diseases in transgenic tobacco. J Exp Bot 60:3781–3796. doi: 10.1093/jxb/erp214

Printed in the United States
By Bookmasters